FRANCIS BACON

THE ADVANCEMENT OF LEARNING

FRANCIS BACON

THE ADVANCEMENT OF LEARNING

Edited by G. W. KITCHIN

Introduction by JERRY WEINBERGER

PAUL DRY BOOKS   *Philadelphia*   2001

First Paul Dry Books Edition, 2001

Paul Dry Books, Inc.
Philadelphia, Pennsylvania
www.pauldrybooks.com

Text type: Minion
Display type: Trajan
Composed by Duke & Company
Designed by Adrianne Onderdonk Dudden

1 3 5 7 9 8 6 4 2
Printed in the United States of America

Library of Congress Cataloging-in-Publication Data

Bacon, Francis, 1561–1626.
    The advancement of learning / Francis Bacon ; edited by G. W. Kitchin ;
introduction by Jerry Weinberger.
        p. cm.
    Originally published: 1861.
    Includes bibliographical references and index.
    ISBN 0-9664913-6-X (pbk. : alk. paper)
    1. Science—Methodology—Early works to 1800. 2. Logic—Early works
to 1800. 3. Utopias—Early works to 1800. I. Kitchin, G. W. (George William),
1827–1912. II. Title.

B1190 .B33 2001
121—dc21
                                                                    00-069389
ISBN 0-9664913-6-X

CONTENTS ❧

INTRODUCTION
BY JERRY WEINBERGER

As fortunes and reputations go, Francis Bacon's have had their ups and downs. Born into an influential family in 1561, upon the death of his father eight years later Bacon was left without the resources of wealth and influence. He eventually became a lawyer and entered Parliament but was frustrated in his search for a place in Queen Elizabeth's court, hampered by his lack of connections and the Queen's ill favor. When Elizabeth died in 1603, Bacon finally began the meteoric rise his talents promised, and by 1618 he had become Lord Chancellor, the highest legal office in England. Soon after, he was made a baron and then, three years later, a viscount. Not many months after reaching the pinnacle, however, Bacon was brought low in one of the most famous scandals in Western political and intellectual history: He was impeached and convicted for having taken bribes in cases that had come before his court. Although King James forgave most of the punishments heaped upon him, Bacon never again played a role in public life and spent his remaining years in relative seclusion, composing many of his most important writings. Not even his death in 1626 washed the stain of scandal from his reputation. More than a century later, Benjamin Franklin, in *Poor Richard Improved,* could still comment that Bacon's prodigious theoretical genius was matched by his moral weakness. Alexander Pope —to whom Franklin refers—called Bacon "the wisest, brightest, meanest of mankind," and Lord Macauley reviled Bacon for his dishonesty and treachery against friends. Even today, accounts of Bacon's life inevitably include details of his fall and bad character.

Despite Bacon's checkered moral reputation, he was revered long after his death as the founder of the modern scientific and technological project. In praising his genius, Franklin commented that Bacon was "justly esteem'd the father of modern experimental philosophy." According to this view, Bacon freed the human intellect from its enthrallment to ancient thought and the sterile legacy of that thought, medieval scholasticism. Bacon redirected men from idle and fruitless speculation to the active and empirical investigation of nature, and he showed how real natural science, as opposed to airy metaphysical systems, would lead to the conquest of nature and the "relief of man's estate" (Book One, V. 11). Many great thinkers of the Enlightenment—the likes of Leibnitz, Rousseau, Kant, and Thomas Jefferson—considered Bacon to be one of the greatest thinkers of all time and a crucial founder of the modern world, in which human life is guided and supported only by reason.

But even in this regard, Bacon's once high reputation faded. As early as the mid-eighteenth century, David Hume argued that Bacon was no real founder of modern science, since he didn't know geometry and was blind to the application of mathematics to the understanding of nature. By the mid-nineteenth century, Hume's view was generally accepted. The epistemologist J. Liebig attacked Bacon's inductive method as fallacious. In more recent times, Sir Karl Popper called Bacon's inductive method a "myth," and even Bacon's admiring biographer Catherine Drinker Bowen admitted that he made no scientific discoveries and that his philosophy of science had been useless in the laboratory. Bacon came to be considered a fine stylist and a prescient futurologist but not an important contributor to the actual development of science, especially not to mathematics, physics, and astronomy. At best, Bacon was a cheerleader, or propagandist, of science. Likewise, his importance in the development of modern politics faded, and he was rarely considered among the great modern political theorists, such as Machiavelli, Hobbes, and Locke. His philosophical and historical writings came to be largely neglected. His *Essays* remained as the primary legacy of a man remembered more for his moral failings than for any contributions to modern science or technological progress.

By the middle of the twentieth century, however, serious questions arose about the contribution of modern science to genuine human progress. Rather than unambiguous forces for good, science and technology had served revolutionary and totalitarian madness such as the world had never seen. Critics questioned the place of technology in human life and argued that technology had become more master than servant. The philosophers Friedrich Nietzsche and Martin Heidegger had pointed out that, while modern science might explain and control nature, it could not advise men how they should use that control or whether they should even pursue such control at all. Nor had science, they argued, really shown that it could explain the whole of experience or that life could be lived according to reason alone. This kind of rationalism was really an expression of faith in—not proof of—the autonomy of reason. At most, modern science could only deafen human beings to the call of the mysterious and the divine. But science had not seriously considered the possibility of such a call. Indeed, science, itself a form of faith, could not rule out the rise of some new religion, which might, one day, replace our faith in science altogether. As a result of this skeptical climate, particularly in recent decades, interest in Bacon as more than a literary and historical figure began to revive. After all, Bacon was the first to address the issues that had again become so pressing in our time: Why should we pursue scientific progress? What are the implications of modern science for religion, faith, and morality? Does technology enhance or disfigure the human soul? Bacon discussed precisely those matters, which have become humanity's most pressing concerns.

While busy with his legal and parliamentary career and hoping for high office, Bacon published *The Advancement of Learning*, in English, in 1605. He dedicated the book to the new King James, who, Bacon knew, was much more disposed to his political advancement than Elizabeth had been, and whom Bacon addressed as the most learned and philosophical ruler since the time of Christ. In calling the King a philosopher, Bacon was engaging in more than flattery. He was boldly inviting the King to join him in heralding a complete revolution in the definition of what knowledge is, what knowledge is for, and how knowledge is to be pursued.

That these are not simply academic matters, Bacon makes clear at the very outset of the treatise, since in Book One he addresses those with the most at stake in any transformation of learning: the "Divines," the "Politiques," and the "learned men themselves." What religious, political, and academic authorities think to be true certainly bears on the power they wield. They have an interest, then, in any difference between what they think is true and what someone else claims is really true. Given their power and authority, they are well placed to resist and punish any reformer of learning who might threaten their positions.

When he directs himself to these three groups, Bacon proceeds as if to defend learning from the cavils of the theologians and the politicians and from the excesses and stupidities of certain of the learned themselves. It seems at first as if Bacon sees the politicians and theologians, on one side, with true learning on another, and bad learning off to the sideline, but the situation he confronts is more complicated. The theologians were obviously involved with learning. In fact, the retardation and the degenerate state of learning was manifested in the scholasticism that Bacon himself had studied at Cambridge and that had repelled him. It sprang from a mixing of philosophy and theology and was hatched, in Bacon's words, by men "shut up in the cells of monasteries and colleges" (Book One, IV. 5). In his dedication to King James, Bacon says that the king is the most learned and philosophical of rulers, but he makes it clear that many rulers have been interested in and influenced by learning. The real condition of the world of his time, Bacon thought, was that a degenerate form of learning—one that stifled genuine science and retarded human progress—bound political power to theological power.

Bacon thus asks the King to support some pretty dangerous, indeed revolutionary, business. No wonder he calls his treatise "The Advancement of Learning," as if it consisted of recommendations for advancing what had already been accomplished, rather than some more accurate title, such as "Why What is Most Important in Existing Learning Must Be Junked and Replaced by an Entirely New Understanding of Physical and Human Nature." In 1620, just before his disgrace and fall from power, and the same year in which he published his other major theoretical treatise, *Novum Organum* ("New Reasoning Instrument"), Bacon wrote *Great Instauration,* a brief introduction to the several theoretical works comprising his plan for human learning. The title, which means "great renewal," again seems to promise an invigoration of what had come before. But in his preliminary remarks, Bacon makes it clear that he is bent on a "total reconstruction of sciences, arts, and all human knowledge, raised upon the proper foundations." *The Advancement of Learning*—and its later expanded version in Latin—presents the basic arguments for this radical departure for all human knowledge.

In a nutshell, Bacon argues that all important learning prior to his complete reconstruction was false because it failed to respect the difference between philosophy, which studies nature, and theology, which studies the divine. This mistake was understandable in the case of the ancient Greek philosophers who, being pagans, believed the entire natural world to be a divine image of God. The mistake should have disappeared with the advent of Christianity, however, which teaches that only humanity exists as the image of God, that the world is merely created, and that nothing created is divine. But the inherent weakness of the human mind permitted the confusion of philosophy and theology to infect Christian thought as well, as evidenced by its tendency to think that nature points to and actually discloses divine providence.

This confusion of philosophy and theology, says Bacon, leads the human mind to a false conception of nature. The world is taken as a conglomeration of things—human beings, trees, mountains, lions, water, etc.—all held together in a coherent and ranked order. Such an order is "teleological," which means that every thing has its specific purpose and thus its own special excellence and that the various things in nature are connected in harmonious motion toward some final end. It makes no difference, according to Bacon, if such a teleological order is described as natural (as with Aristotle) or as divine (as with Plato and corrupted Christian thought). The former falsely conceives of nature as if it were God; the latter falsely conceives of nature as a god-like extension of God.

To describe the effects of such thinking, Bacon uses the example of an eyelash (Book Two, VII. 7). When one thinks teleologically, says Bacon, one is satisfied simply to point out that eyelashes conduce to sight by keeping dirt out of the eyes. Eyelashes are for sight and sight serves many human purposes, the most important and highest of which is to behold the beauty of God's creation. Knowing these facts, we can supposedly distinguish between good and bad eyelashes. Moreover, since the natural order is divine, it cannot be transient, as one might conclude when one observes that individual eyelashes and human beings and trees and lions die and pass out of existence. Therefore, the "true" world—the world that does not change—consists of the immaterial *forms* of the particular things that, being material, come and go. There is thus a pure form of eyelash and lion and human being, and these pure forms and their relations to one another are the highest objects of knowledge and the ultimate reality of nature.

Bacon completely rejects the teleological vision of nature, which he thought had not only persisted despite the advent of Christianity, but had in fact been absorbed into Christian thought and learning. For Bacon, the teleological approach to nature was mere circular reasoning and led absolutely nowhere, except to the convoluted gibberish of scholastic metaphysics. Since it focuses on spurious forms and nonexistent relations among things, it ignores what really occurs in nature: the many and complex material motions that cause what happens around us, but that most often we cannot see by our unaided eyes and minds. To say that eyelashes are for sight is merely to state the obvious. Genuine science inquires into the material causes: for example, damp organs, such as eyes, are protected by surrounding, and different tissues, such as lashes. The patterns of these material causes are the real forms of nature and can be revealed by using the right experimental method. This method puts nature on an experimental rack. It pulls and pushes and hammers at nature, in order to force nature to reveal its often hidden material causes. When we know these causes, we can change nature, whether visible or not to the human eye, and this power to change nature can be put to human use for the relief of man's estate. Moreover, by changing nature in this way we reveal what nature really is. Thus, there is in Bacon's view a perfect harmony between making new things and discovering what is true about nature.

Today, we live in a world transformed by science and technology, just as Bacon envisioned it could be. True, we would not learn from Bacon's writings much about the actual sciences and experimental methods that have fashioned our world. But by confronting as he did the authorities, who stood to lose a great deal with the advent of an entirely new knowledge, Bacon was forced to address questions that for centuries, at least until recently, were mostly taken for granted. Suppose it is true that the old, teleological conception of nature is false and that things have no special purposes or ends. If so, then does it follow that there is no special, natural virtue or end for human beings? Is reason powerless to discover what the best human life is, or what moral rules should govern the human will? If such questions cannot be answered by science, then one might argue that mankind confronts two fundamental choices: either faith guides it along its way or mankind must posit, arbitrarily, its own path. Indeed, Bacon suggests that, as Christianity frees the mind from the pagan superstition that nature is divine, and thus makes natural science possible, so natural science clears the air for considering the true laws of God. But suppose the comforts afforded by natural science cause faith to weaken or disappear? How then will we live with the powers that science and technology place in our hands? Bacon could not propose his revolution in learning without considering, as delicately as he could, these and many similar questions.

And consider them he did, in the array of writings on philosophy, science, politics, history, and morality that he bequeathed to the world he heralded. *The Advancement of Learning* was the first of these writings. In it, Bacon shows us the forces arrayed against modern science and then surveys the entire landscape of existing human knowledge. He intends to show what is bad and good in existing knowledge, to show what is missing altogether, and to suggest remedies. In presenting this comprehensive picture, Bacon describes the arts and sciences as an organic unity. They "are not like several lines that meet in one angle, and so touch but in a point: but are like branches of a tree, that meet in a stem, which hath a dimension and quantity of entireness and continuance, before it come to discontinue and break itself into arms and boughs" (Book Two, V. 2). The various forms of knowledge—from physics and medicine to politics, history, and morality—are all related. It is impossible for a change in one important branch of knowledge not to bear on all the others. Bacon knew two things about his rejection of teleology. First, this change represented a radical challenge to existing academic, political, and ecclesiastical establishments. Second, it would have a significant effect on our commonsensical assumptions about what is good and bad, noble and base, just and unjust.

Bacon's appreciation of the unity of knowledge—now often forgotten in the fragmented world of modern specialization—enabled him to question, in a profound and eloquent way, the ultimate value of science and technological progress. He invites the reader to consider the relations among modern science and technology, on the one hand, and politics, religion, and morality, on the other. It is hard to imagine a book more attuned to our times. In his personal life, Bacon may have been a scoundrel, and perhaps deserving of the scorn of later generations. However, in the spirit of the science he did so much to promote, we can relish what his brilliant mind and pen offer us, especially since we need it so much now. In this spirit, *The Advancement of Learning*—as edited by G. W. Kitchin in 1861—is again offered to the reading public.

A NOTE ON THE TEXT

The primary source for this edition of *The Advancement of Learning* is G. W. Kitchin's 1861 text, supplemented by W. A. Wright's 1868 work. The paragraph-numbering scheme is Wright's.

Editor's note on the text
In preparing this edition of the *Advancement of Learning* for the general reader, I have aimed at three things—a faithful text, full verification of quotations, and brevity and simplicity of notes.

As to the first of these matters, there was but little difficulty. The variations in the text are very few, and very unimportant. Wherever it was possible, I have followed the edition of 1605, leaving myself little scope for conjecture.

As to the next point, I had the work already done for me, to a great extent, both in the edition of Mr Markby, and in the *De Augmentis* of the great Ellis and Spedding edition. I have been able here and there to supply missing references, and have carefully verified those already found for me.

But with respect to notes, it is unnecessary that I say more than that their aim is to be as unobtrusive as possible, and that I hope they may be useful.

Lastly, I subjoin a brief analysis of the work.

BOOK I. (*Preliminary.*) Briefly removes the prejudices against Learning, with proofs, divine and human, of its dignity. (Corresponds with *De Augmentis*, Bk. I.) *3–57*

BOOK II. (*On the main subject.*) Commended to kings as nursing fathers. (*De Augm. ii. proef.*) *61–68*

Learning is twofold—Divine and Human. Divine postponed. (*De Augm.* ii). *83, 87*

Human Learning is threefold—I. History (which answers to the *Memory*). II. Poesy (*to Imagination*). III. Philosophy (*to Reason*). *68*

G.W.K.

THE PHILOSOPHY TREE

A Representation of the Divisions of Philosophy Proposed by Francis Bacon (Book Two, pp. 81–190)

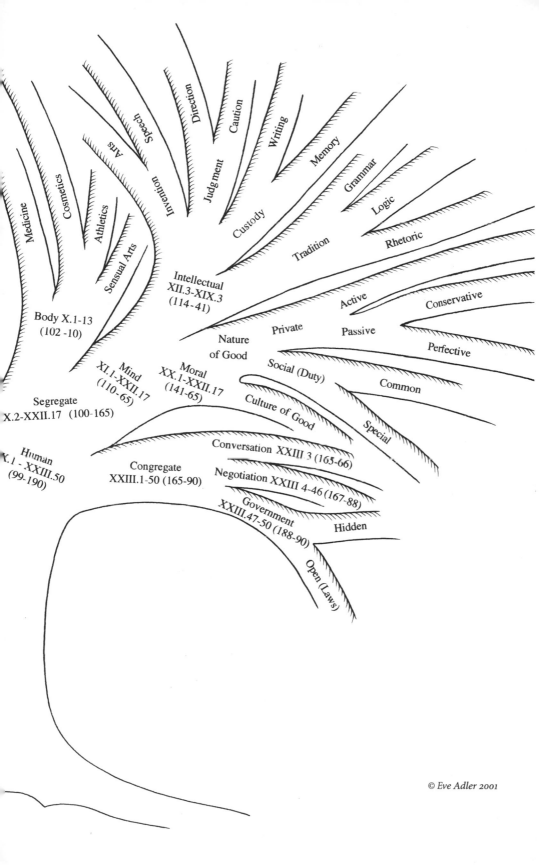

Medicine

Cosmetics

Athletics

Sensual Arts

Arts

Invention

Speech

Direction

Caution

Judgment

Writing

Memory

Custody

Grammar

Logic

Tradition

Rhetoric

Body X.1-13
(102 -10)

Intellectual
XII.3-XIX.3
(114-41)

Active

Conservative

Private

Passive

Perfective

Nature
of Good

Mind
XI.1-XXII.17
(110-65)

Moral
XX.1-XXII.17
(141-65)

Social (Duty)

Common

Segregate
X.2-XXII.17 (100-165)

Culture of Good

Special

Conversation XXIII 3 (165-66)

Human
X.1 - XXIII.50
(99-190)

Congregate
XXIII.1-50 (165-90)

Negotiation XXIII 4-46 (167-88)

Government
XXIII.47-50 (188-90)

Hidden

Open (Laws)

© Eve Adler 2001

THE FIRST BOOK

THE FIRST BOOK OF FRANCIS BACON
OF THE PROFICIENCE AND ADVANCEMENT
OF LEARNING DIVINE AND HUMAN

To the King

1. There were under the law, excellent King, both daily Sacrifices and free-will offerings; the one proceeding upon ordinary Observance, the other upon a devout cheerfulness: in like manner there belongeth to Kings from their servants both tribute of duty and presents of affection. In the former of these I hope I shall not live to be wanting, according to my most humble duty, and the good pleasure of your Majesty's employments: for the latter, I thought it more respective to make choice of some oblation, which might rather refer to the propriety and excellency of your individual person, than to the business of your crown and state.

2. Wherefore, representing your Majesty many times unto my mind, and beholding you not with the inquisitive eye of presumption, to discover that which the Scripture telleth me is inscrutable,[1] but with the observant eye of duty and admiration; leaving aside the other parts of your virtue and fortune, I have been touched, yea, and possessed with an extreme wonder at those your virtues and faculties, which the Philosophers call intellectual; the largeness of your capacity, the faithfulness of your memory, the swiftness of your apprehension, the penetration of your judgment, and the facility and order of your elocution: and I have often thought that of all the persons living that I have known, your Majesty were the best instance to make a man of Plato's opinion,[2] that all knowledge is but remembrance, and that the mind of man by nature knoweth all things, and hath but her own native and original notions (which by the strangeness and darkness of this tabernacle of the body are sequestered) again revived and restored: such a light of nature I have observed in your Majesty, and such a readiness to take

flame and blaze from the least occasion presented, or the least spark of another's knowledge delivered. And as the Scripture saith of the wisest king, *That his heart was as the sands of the sea;*[3] which though it be one of the largest bodies, yet it consisteth of the smallest and finest portions; so hath God given your Majesty a composition of understanding admirable, being able to compass and comprehend the greatest matters, and nevertheless to touch and apprehend the least; whereas it should seem an impossibility in nature for the same instrument to make itself fit for great and small works. And for your gift of speech, I call to mind what Cornelius Tacitus saith of Augustus Caesar: *Augusto profluens, et quae principem deceret, eloquentia fuit.*[4] For if we note it well, speech that is uttered with labour and difficulty, or speech that savoureth of the affectation of art and precepts, or speech that is framed after the imitation of some pattern of eloquence, though never so excellent; all this hath somewhat servile, and holding of the subject. But your Majesty's manner of speech is indeed prince-like, flowing as from a fountain, and yet streaming and branching itself into nature's order, full of facility and felicity, imitating none, and inimitable by any. And as in your civil estate there appeareth to be an emulation and contention of your Majesty's virtue with your fortune; a virtuous disposition with a fortunate regiment; a virtuous expectation (when time was) of your greater fortune, with a prosperous possession thereof in the due time; a virtuous observation of the laws of marriage, with most blessed and happy fruit of marriage; a virtuous and most Christian desire of peace, with a fortunate inclination in your neighbour princes thereunto: so likewise in these intellectual matters, there seemeth to be no less contention between the excellency of your Majesty's gifts of nature and the universality and perfection of your learning. For I am well assured that this which I shall say is no amplification at all, but a positive and measured truth; which is, that there hath not been since Christ's time any King or temporal Monarch, which hath been so learned in all literature and erudition, divine and human. For let a man seriously and diligently revolve and peruse the succession of the emperors of Rome; of which Caesar the Dictator, who lived some years before Christ, and Marcus Antoninus were the best learned; and so descend to the emperors of Graecia, or of the West, and then to the lines of France,

Spain, England, Scotland, and the rest, and he shall find this judgment is truly made. For it seemeth much in a King, if, by the compendious extractions of other men's wits and labours, he can take hold of any superficial ornaments and shows of learning; or if he countenance and prefer learning and learned men: but to drink indeed of the true fountains of learning, nay, to have such a fountain of learning in himself, in a King, and in a King born, is almost a miracle. And the more, because there is met in your Majesty a rare conjunction, as well of divine and sacred literature, as of profane and human; so as your Majesty standeth invested of that triplicity, which in great veneration was ascribed to the ancient Hermes; the power and fortune of a king, the knowledge and illumination of a priest, and the learning and universality of a philosopher.[5] This propriety inherent[6] and individual attribute in your Majesty deserveth to be expressed not only in the fame and admiration of the present time, nor in the history or tradition of the ages succeeding, but also in some solid work, fixed memorial, and immortal monument, bearing a character or signature both of the power of a King, and the difference and perfection of such a King.

3.   Therefore I did conclude with myself, that I could not make unto your Majesty a better oblation than of some Treatise tending to that end, whereof the sum will consist of these two parts; the former, concerning the excellency of Learning and Knowledge, and the excellency of the merit and true glory in the augmentation and propagation thereof: the latter, what the particular acts and works are, which have been embraced and undertaken for the Advancement of Learning; and again, what defects and undervalues I find in such particular acts: to the end, that though I cannot positively or affirmatively advise your Majesty, or propound unto you framed particulars; yet I may excite your princely cogitations to visit the excellent treasure of your own mind, and thence to extract particulars for this purpose, agreeably to your magnanimity and wisdom.

I   1.   In the entrance to the former of these, to clear the way, and as it were to make silence, to have the true testimonies concerning the dignity of Learning to be better heard, without the interruption of tacit objections, I think good to deliver it from the discredits and disgraces which it hath received, all from ignorance; but ignorance severally disguised, appearing sometimes in the zeal and jealousy of Divines; sometimes in the severity and arrogancy of Politiques; and sometimes in the errors and imperfections of learned men themselves.

2.   I hear the former sort say, that Knowledge is of those things which are to be accepted of with great limitation and caution; that the aspiring to overmuch knowledge was the original temptation and sin whereupon ensued the fall of man; that Knowledge hath in it somewhat of the serpent, and therefore where it entereth into a man it makes him swell; *Scientia inflat:*[7] that Salomon gives a censure, *That there is no end of making books, and that much reading is weariness of the flesh;*[8] and again in another place, *That in spacious knowledge there is much contristation, and that he that increaseth knowledge increaseth anxiety;*[9] that Saint Paul gives a caveat, *That we be not spoiled through vain philosophy;*[10] that experience demonstrates how learned men have been archheretics, how learned times have been inclined to atheism, and how the contemplation of second causes doth derogate from our dependence upon God, who is the first cause.

3.   To discover then the ignorance and error of this opinion, and the misunderstanding in the grounds thereof, it may well appear these men do not observe or consider that it was not the pure knowledge of nature and universality, a knowledge by the light whereof man did give names unto other creatures in Paradise,[11] as they were brought before him, according unto their proprieties, which gave the occasion to the fall: but it was the proud knowledge of good and evil, with an intent in man to give law unto himself, and to depend no more upon God's commandments, which was the form of the temptation. Neither is it any quantity of knowledge, how great soever, that can make the mind of man to swell; for nothing can fill, much less extend the soul of man, but God and the contemplation of God; and therefore Salomon, speaking of the two principal senses of inquisition, the eye and the ear, affirmeth that the eye is never satisfied with seeing, nor the ear with hearing;[12] and if there be no fulness, then is the continent greater than the content: so of knowledge itself, and the mind of man, whereto the senses are but reporters, he defineth likewise in these words, placed

after that Kalendar or Ephemerides which he maketh of the diversities of times and seasons for all actions and purposes; and concludeth thus: *God hath made all things beautiful, or decent, in the true return of their seasons: Also he hath placed the world in man's heart, yet cannot man find out the work which God worketh from the beginning to the end:*[13] declaring not obscurely, that God hath framed the mind of man as a mirror or glass, capable of the image of the universal world, and joyful to receive the impression thereof, as the eye joyeth to receive light; and not only delighted in beholding the variety of things and vicissitude of times, but raised also to find out and discern the ordinances and decrees, which throughout all those changes are infallibly observed. And although he doth insinuate that the supreme or summary law of nature, which he calleth, *The work which God worketh from the beginning to the end,* is not possible to be found out by man; yet that doth not derogate from the capacity of the mind, but may be referred to the impediments, as of shortness of life, ill conjunction of labours, ill tradition of knowledge over from hand to hand, and many other inconveniences, whereunto the condition of man is subject. For that nothing parcel of the world is denied to man's inquiry and invention, he doth in another place rule over, when he saith, *The spirit of man is as the lamp of God, wherewith he searcheth the inwardness of all secrets.*[14] If then such be the capacity and receipt of the mind of man, it is manifest that there is no danger at all in the proportion or quantity of knowledge, how large soever, lest it should make it swell or outcompass itself; no, but it is merely the quality of knowledge, which, be it in quantity more or less, if it be taken without the true corrective thereof, hath in it some nature of venom or malignity, and some effects of that venom, which is ventosity or swelling. This corrective spice, the mixture whereof maketh Knowledge so sovereign, is Charity, which the Apostle immediately addeth to the former clause: for so he saith, *Knowledge bloweth up, but Charity buildeth up;* not unlike unto that which he delivereth in another place: *If I spake,* saith he, *with the tongues of men and angels, and had not charity, it were but as a tinkling cymbal;*[15] not but that it is an excellent thing to speak with the tongues of men and angels, but because, if it be severed from charity, and not referred to the good of men and mankind, it hath rather a sounding and unworthy glory, than a meriting and substantial virtue. And as for that censure of Salomon, concerning the excess of writing and reading books, and the anxiety of spirit which redoundeth from knowledge; and that admonition of Saint Paul, *That we be not seduced by vain philosophy;* let those places be rightly understood, and they do indeed excellently set forth the true bounds and limitations, whereby human knowledge is confined and circumscribed; and yet without any such contracting or coarctation, but that it may comprehend all the universal nature of things; for these limitations are three: the first, *That we do not so place our felicity in knowledge, as we forget our mortality:* the second, *That we make application of our knowledge, to give ourselves repose and contentment, and not distaste or repining:* the third, *That we do not presume by the contemplation of nature to attain to the mysteries of God.* For as touching the first of these, Salomon doth excellently expound himself in another place of the same book, where he saith:[16] *I saw well that knowledge recedeth as far from ignorance as light doth from darkness; and that the wise man's eyes keep watch in his head, whereas the fool roundeth about in darkness: but withal I learned, that the same mortality involveth them both.* And for the second, certain it is, there is no vexation or anxiety of mind which resulteth from knowledge otherwise than merely by accident; for all knowledge and wonder (which is the seed of knowledge) is an impression of pleasure in itself: but when men fall to framing conclusions out of their knowledge, applying it to their particular, and ministering to themselves thereby weak fears or vast desires, there groweth that carefulness and trouble of mind which is spoken of: for then knowledge is no more *Lumen siccum,* whereof Heraclitus the profound[17] said, *Lumen siccum optima anima;* but it becometh *Lumen madidum,* or *maceratum,* being steeped and infused in the humours of the affections.[18] And as for the third point, it deserveth to be a little stood upon, and not to be lightly passed over: for if any man shall think by view and inquiry into these sensible and material things to attain that light, whereby he may reveal unto himself the Nature or Will of God, then indeed is he spoiled by vain philosophy: for the contemplation of God's creatures and works produceth (having regard to the works and creatures themselves) knowledge, but having regard to God, no perfect knowledge, but wonder, which is broken knowledge. And therefore it was most aptly said by one of Plato's school,[19] *That the sense of man carrieth a resemblance with the sun, which, as we see, openeth and revealeth all the terrestrial globe; but then again it obscureth and concealeth the stars and celestial globe: so doth the sense discover natural things, but it darkeneth and shutteth up divine.* And hence it is true that it hath proceeded, that divers great learned men have been heretical, whilst they have sought to fly up to the secrets of the Deity by the waxen wings of the senses. And as for the conceit that too much knowledge should incline a man to Atheism,[20] and that the ignorance of second causes should make a more devout dependence upon God, which is the first cause; first, it is good to ask the question which Job asked of his friends: *Will you lie for God, as one man will do for another, to gratify him?*[21] For certain it is that God worketh nothing in nature but by second causes: and if they would have it otherwise believed, it is mere imposture, as it were in favour towards God; and nothing else but to offer to the Author of Truth the unclean sacrifice of a lie. But further, it is an assured truth, and a conclusion of experience, that a little or superficial knowledge of Philosophy may incline the mind of man to Atheism, but a further proceeding therein doth bring the mind back again to Religion: for in the entrance of Philosophy, when the second causes, which are next unto the senses, do offer themselves to the mind of man, if it dwell and stay there it may induce some oblivion of the highest cause; but when a man passeth on further, and seeth the dependence of causes, and the works of Providence, then, according to the allegory of the poets, he will easily believe that the highest link of nature's chain must needs be tied to the foot of Jupiter's chair.[22] To conclude therefore, let no man upon a weak conceit of sobriety or an ill-applied moderation think or maintain, that a man can search too far, or be too well studied in the book of God's word, or in the book of God's works, divinity or philosophy; but rather let men endeavour an endless progress or proficience in both; only let men beware that they apply both to charity, and not to swelling; to use, and not to ostentation; and again, that they do not unwisely mingle or confound these learnings together.

II 1. And as for the disgraces which Learning receiveth from Politiques, they be of this nature; that Learning doth soften men's minds, and makes them more unapt for the honour and exercise of arms; that it doth mar and pervert men's dispositions for matter of government and policy, in making them too curious and irresolute by variety of reading, or too peremptory or positive by strictness of rules and axioms, or too immoderate and overweening by reason of the greatness of examples, or too incompatible and differing from the times by reason of the dissimilitude of examples; or at least, that it doth divert men's travails from action and business, and bringeth them to a love of leisure and privateness; and that it doth bring into states a relaxation of discipline, whilst every man is more ready to argue than to obey and execute. Out of this conceit, Cato,[23] surnamed the Censor, one of the wisest men indeed that ever lived, when Carneades the philosopher came in embassage to Rome, and that the young men of Rome began to flock about him, being allured with the sweetness and majesty of his eloquence and learning, gave counsel in open senate that they should give him his dispatch with all speed, lest he should infect and enchant the minds and affections of the youth, and at unawares bring in an alteration of the manners and customs of the state.[24] Out of the same conceit or humour did Virgil, turning his pen to the advantage of his country, and the disadvantage of his own profession, make a kind of separation between policy and government, and between arts and sciences, in the verses so much renowned, attributing and challenging the one to the Romans and leaving and yielding the other to the Grecians: *Tu regere imperio populos, Romane, memento, Hae tibi erunt artes,* etc.[25] So likewise we see that Anytus, the accuser of Socrates, laid it as an article of charge and accusation against him, that he did, with the variety and power of his discourses and disputations, withdraw young men from due reverence to the laws and customs of their country, and that he did profess a dangerous and pernicious science, which was, to make the worse matter seem the better, and to suppress truth by force of eloquence and speech.[26]

2. But these and the like imputations have rather a countenance of gravity than any ground of justice: for experience doth warrant, that

both in persons and in times, there hath been a meeting and concurrence in Learning and Arms, flourishing and excelling in the same men and the same ages. For as for men, there cannot be a better nor the like instance, as of that pair, Alexander the Great and Julius Caesar the Dictator; whereof the one was Aristotle's scholar in philosophy, and the other was Cicero's rival in eloquence: or if any man had rather call for scholars that were great generals, than generals that were great scholars, let him take Epaminondas the Theban, or Xenophon the Athenian; whereof the one was the first that abated the power of Sparta, and the other was the first that made way to the overthrow of the monarchy of Persia. And this concurrence is yet more visible in times than in persons, by how much an age is a greater object than a man. For both in Egypt, Assyria, Persia, Graecia, and Rome, the same times that are most renowned for arms, are likewise most admired for learning, so that the greatest authors and philosophers, and the greatest captains and governors have lived in the same ages. Neither can it otherwise be: for as in man the ripeness of strength of the body and mind cometh much about an age, save that the strength of the body cometh somewhat the more early:[27] so in states, Arms and Learning, whereof the one correspondeth to the body, the other to the soul of man, have a concurrence or near sequence in times.

3.　And for matter of Policy and Government, that learning should rather hurt, than enable thereunto, is a thing very improbable: we see it is accounted an error to commit a natural body to empiric physicians, which commonly have a few pleasing receipts whereupon they are confident and adventurous, but know neither the causes of diseases, nor the complexions of patients, nor peril of accidents, nor the true method of cures: we see it is a like error to rely upon advocates or lawyers, which are only men of practice and not grounded in their books, who are many times easily surprised when matter falleth out besides their experience, to the prejudice of the causes they handle: so by like reason it cannot be but a matter of doubtful consequence if states be managed by empiric Statesmen, not well mingled with men grounded in learning. But contrariwise, it is almost without instance contradictory that ever any government was disastrous that was in the hands of learned governors.[28] For howsoever it hath been ordinary with politique men to extenuate and disable learned men by the names of *Pedantes;* yet in the records of time it appeareth in many particulars that the governments of princes in minority (notwithstanding the infinite disadvantage of that kind of state) have nevertheless excelled the government of princes of mature age, even for that reason which they seek to traduce, which is, that by that occasion the state hath been in the hands of *Pedantes:* for so was the state of Rome for the first five years, which are so much magnified, during the minority of Nero, in the hands of Seneca, a *Pedanti;* so it was again, for ten years' space or more, during the minority of Gordianus the younger, with great applause and contentation in the hands of Misitheus, a *Pedanti:* so was it before that, in the minority of Alexander Severus, in like happiness, in hands not much unlike, by reason of the rule of the women, who were aided by the teachers and preceptors. Nay, let a man look into the government of the bishops of Rome, as by name, into the government of Pius Quintus and Sextus Quintus in our times, who were both at their entrance esteemed but as pedantical friars, and he shall find that such popes do greater things, and proceed upon truer principles of estate, than those which have ascended to the papacy from an education and breeding in affairs of estate and courts of princes; for although men bred in learning are perhaps to seek in points of convenience and accommodating for the present, which the Italians call *Ragioni di stato,* whereof the same Pius Quintus could not hear spoken with patience, terming them inventions against religion and the moral virtues; yet on the other side, to recompense that, they are perfect in those same plain grounds of religion, justice, honour, and moral virtue, which if they be well and watchfully pursued, there will be seldom use of those other, no more than of physic in a sound or well-dieted body. Neither can the experience of one man's life furnish examples and precedents for the events of one man's life: for as it happeneth sometimes that the grandchild, or other descendant, resembleth the ancestor more than the son; so many times occurrences of present times may sort better with ancient examples than with those of the later or immediate times; and lastly, the wit of one man can no more countervail learning than one man's means can hold way with a common purse.

4.   And as for those particular seducements or indispositions of the mind for policy and government, which Learning is pretended to insinuate; if it be granted that any such thing be, it must be remembered withal, that Learning ministereth in every of them greater strength of medicine or remedy than it offereth cause of indisposition or infirmity. For if by a secret operation it make men perplexed and irresolute, on the other side by plain precept it teacheth them when and upon what ground to resolve; yea, and how to carry things in suspense without prejudice, till they resolve; if it make men positive and regular, it teacheth them what things are in their nature demonstrative, and what are conjectural, and as well the use of distinctions and exceptions, as the latitude of principles and rules. If it mislead by disproportion or dissimilitude of examples, it teacheth men the force of circumstances, the errors of comparisons, and all the cautions of application; so that in all these it doth rectify more effectually than it can pervert. And these medicines it conveyeth into men's minds much more forcibly by the quickness and penetration of examples. For let a man look into the errors of Clement the seventh, so lively described by Guicciardine,[29] who served under him, or into the errors of Cicero, painted out by his own pencil in his Epistles to Atticus, and he will fly apace from being irresolute. Let him look into the errors of Phocion, and he will beware how he be obstinate or inflexible. Let him but read the fable of Ixion,[30] and it will hold him from being vaporous or imaginative. Let him look into the errors of Cato the second, and he will never be one of the *Antipodes,* to tread opposite to the present world.[31]

5.   And for the conceit that Learning should dispose men to leisure and privateness, and make men slothful; it were a strange thing if that which accustometh the mind to a perpetual motion and agitation should induce slothfulness: whereas contrariwise it may be truly affirmed, that no kind of men love business for itself but those that are learned; for other persons love it for profit, as a hireling, that loves the work for the wages; or for honour, as because it beareth them up in the eyes of men, and refresheth their reputation, which otherwise would wear; or because it putteth them in mind of their fortune, and giveth them occasion to pleasure and displeasure; or because it exerciseth some faculty wherein they take pride, and so entertaineth them in good

humour and pleasing conceits towards themselves; or because it advanceth any other their ends. So that as it is said of untrue valours, that some men's valours are in the eyes of them that look on; so such men's industries are in the eyes of others, or at least in regard of their own designments: only learned men love business as an action according to nature, as agreeable to health of mind as exercise is to health of body, taking pleasure in the action itself, and not in the purchase: so that of all men they are the most indefatigable, if it be towards any business which can hold or detain their mind.

6.   And if any man be laborious in reading and study and yet idle in business and action, it groweth from some weakness of body or softness of spirit; such as Seneca speaketh of: *Quidam tam sunt umbratiles, ut putent in turbido esse quicquid in luce est;*[32] and not of Learning: well may it be that such a point of a man's nature may make him give himself to Learning, but it is not learning that breedeth any such point in his nature.

7.   And that Learning should take up too much time or leisure, I answer, the most active or busy man that hath been or can be, hath, no question, many vacant times of leisure, while he expecteth the tides and returns of business (except he be either tedious and of no dispatch, or lightly and unworthily ambitious to meddle in things that may be better done by others), and then the question is but how those spaces and times of leisure shall be filled and spent; whether in pleasures or in studies; as was well answered by Demosthenes to his adversary Aeschines, that was a man given to pleasure, and told him, *That his orations did smell of the lamp: Indeed* (said Demosthenes) *there is a great difference between the things that you and I do by lamp-light.*[33] So as no man need doubt that learning will expulse business, but rather it will keep and defend the possession of the mind against idleness and pleasure, which otherwise at unawares may enter to the prejudice of both.

8.   Again, for that other conceit that Learning should undermine the reverence of laws and government, it is assuredly a mere depravation and calumny, without all shadow of truth. For to say that a blind custom of obedience should be a surer obligation than duty taught and understood, it is to affirm, that a blind man may tread surer by a guide than a seeing man can by a light. And it is without all controversy, that

learning doth make the minds of men gentle, generous, maniable, and pliant to government; whereas ignorance makes them churlish, thwart, and mutinous: and the evidence of time doth clear this assertion, considering that the most barbarous, rude, and unlearned times have been most subject to tumults, seditions, and changes.

9.   And as to the judgment of Cato the Censor, he was well punished for his blasphemy against Learning, in the same kind wherein he offended; for when he was past threescore years old, he was taken with an extreme desire to go to school again, and to learn the Greek tongue, to the end to peruse the Greek authors; which doth well demonstrate that his former censure of the Grecian learning was rather an affected gravity, than according to the inward sense of his own opinion. And as for Virgil's verses, though it pleased him to brave the world in taking to the Romans the art of empire, and leaving to others the arts of subjects; yet so much is manifest that the Romans never ascended to that height of empire, till the time they had ascended to the height of other arts. For in the time of the two first Caesars, which had the art of government in greatest perfection, there lived the best poet, Virgilius Maro; the best historiographer, Titus Livius; the best antiquary, Marcus Varro; and the best, or second orator, Marcus Cicero, that to the memory of man are known. As for the accusation of Socrates, the time must be remembered when it was prosecuted; which was under the Thirty Tyrants, the most base, bloody, and envious persons that have governed; which revolution of state was no sooner over, but Socrates, whom they had made a person criminal, was made a person heroical, and his memory accumulate with honours divine and human; and those discourses of his which were then termed corrupting of manners, were after acknowledged for sovereign medicines of the mind and manners, and so have been received ever since till this day. Let this therefore serve for answer to Politiques, which in their humorous severity, or in their feigned gravity, have presumed to throw imputations upon Learning; which redargution nevertheless (save that we know not whether our labours may extend to other ages) were not needful for the present, in regard of the love and reverence towards Learning, which the example and countenance of two so learned Princes, Queen Elizabeth, and your Majesty, being as Castor and Pollux, *lucida*

sidera,[34] stars of excellent light and most benign influence, hath wrought in all men of place and authority in our nation.

III

1.   Now therefore we come to that third sort of discredit or diminution of credit that groweth unto Learning from learned men themselves, which commonly cleaveth fastest: it is either from their fortune, or from their manners, or from the nature of their studies. For the first, it is not in their power; and the second is accidental; the third only is proper to be handled: but because we are not in hand with true measure, but with popular estimation and conceit, it is not amiss to speak somewhat of the two former. The derogations therefore which grow to Learning from the fortune or condition of learned men, are either in respect of scarcity of means, or in respect of privateness of life and meanness of employments.

2.   Concerning want, and that it is the case of learned men usually to begin with little, and not to grow rich so fast as other men by reason they convert not their labours chiefly to lucre and increase: It were good to leave the common place in commendation of poverty to some friar to handle, to whom much was attributed by Machiavel in this point; when he said, *That the kingdom of the clergy had been long before at an end, if the reputation and reverence towards the poverty of friars had not borne out the scandal of the superfluities and excesses of bishops and prelates.*[35] So a man might say that the felicity and delicacy of princes and great persons had long since turned to rudeness and barbarism, if the poverty of Learning had not kept up civility and honour of life: but without any such advantages, it is worthy the observation what a reverent and honoured thing poverty of fortune was for some ages in the Roman state, which nevertheless was a state without paradoxes. For we see what Titus Livius saith in his introduction: *Caeterum aut me amor negotii suscepti fallit, aut nulla unquam respublica nec major, nec sanctior, nec bonis exemplis ditior fuit; nec in quam tam serae avaritia luxuriaque immigraverint; nec ubi tantus ac tam diu paupertati ac parsimoniae honos fuerit.*[36] We see likewise, after that the state of Rome was not itself, but did degenerate, how that person that took

upon him to be counsellor to Julius Caesar after his victory where to begin his restoration of the state, maketh it of all points the most summary to take away the estimation of wealth: *Verum haec et omnia mala pariter cum honore pecuniae desinent; si neque magistratus, neque alia vulgo cupienda, venalia erunt.*[37] To conclude this point, as it was truly said, that *Rubor est virtutis color,* though sometime it come from vice;[38] so it may be fitly said that *Paupertas est virtutis fortuna,* though sometime it may proceed from misgovernment and accident. Surely Salomon hath pronounced it both in censure, *Qui festinat ad divitias non erit insons;*[39] and in precept, *Buy the truth, and sell it not; and so of wisdom and knowledge;*[40] judging that means were to be spent upon Learning, and not Learning to be applied to means. And as for the privateness, or obscureness (as it may be in vulgar estimation accounted) of life of contemplative men; it is a theme so common to extol a private life, not taxed with sensuality and sloth, in comparison [with] and to the disadvantage of a civil life, for safety, liberty, pleasure, and dignity, or at least freedom from indignity, as no man handleth it but handleth it well; such a consonancy it hath to men's conceits in the expressing, and to men's consents in the allowing. This only I will add, that learned men forgotten in states and not living in the eyes of men, are like images of Cassius and Brutus in the funeral of Junia: of which not being represented, as many others were, Tacitus saith, *Eo ipso praefulgebant, quod non visebantur.*[41]

3.   And for meanness of employment, that which is most traduced to contempt is that the government of youth is commonly allotted to them; which age, because it is the age of least authority, it is transferred to the disesteeming of those employments wherein youth is conversant, and which are conversant about youth. But how unjust this traducement is (if you will reduce things from popularity of opinion to measure of reason) may appear in that we see men are more curious what they put into a new vessel than into a vessel seasoned; and what mould they lay about a young plant than about a plant corroborate; so as the weakest terms and times of all things use to have the best applications and helps. And will you hearken to the Hebrew rabbins? *Your young men shall see visions, and your old men shall dream dreams;*[42] say they youth is the worthier age, for that visions are nearer appa-

ritions of God than dreams? And let it be noted, that howsoever the conditions of life of *Pedantes* have been scorned upon theatres, as the ape of tyranny; and that the modern looseness or negligence hath taken no due regard to the choice of schoolmasters and tutors; yet the ancient wisdom of the best times did always make a just complaint, that states were too busy with their laws and too negligent in point of education: which excellent part of ancient discipline hath been in some sort revived of late times by the colleges of the Jesuits; of whom, although in regard of their superstition I may say, *Quo meliores, eo deteriores;* yet in regard of this, and some other points concerning human learning and moral matters, I may say, as Agesilaus said to his enemy Pharnabazus, *Talis quum sis, utinam noster esses.*[43] And thus much touching the discredits drawn from the fortunes of learned men.

4.   As touching the manners of learned men, it is a thing personal and individual: and no doubt there be amongst them, as in other professions, of all temperatures: but yet so as it is not without truth, which is said, that *Abeunt studia in mores,*[44] studies have an influence and operation upon the manners of those that are conversant in them.

5.   But upon an attentive and indifferent review, I for my part cannot find any disgrace to Learning can proceed from the manners of learned men not inherent[45] to them as they are learned; except it be a fault (which was the supposed fault of Demosthenes, Cicero, Cato the second, Seneca, and many more) that because the times they read of are commonly better than the times they live in, and the duties taught better than the duties practised, they contend sometimes too far to bring things to perfection, and to reduce the corruption of manners to honesty of precepts, or examples of too great height. And yet hereof they have caveats enough in their own walks. For Solon, when he was asked whether he had given his citizens the best laws, answered wisely, *Yea of such as they would receive:*[46] and Plato, finding that his own heart could not agree with the corrupt manners of his country, refused to bear place or office, saying, *That a man's country was to be used as his parents were, that is, with humble persuasions, and not with contestations.*[47] And Caesar's counsellor put in the same caveat, *Non ad vetera instituta revocans quae jampridem corruptis moribus ludibrio sunt;*[48] and Cicero noteth this error directly in Cato the second, when he writes

to his friend Atticus; *Cato optime sentit, sed nocet interdum reipubli-cae; loquitur enim tanquam in reipublica Platonis, non tanquam in faece Romuli.*[49] And the same Cicero doth excuse and expound the philoso-phers for going too far and being too exact in their prescripts, when he saith, *Isti ipsi praeceptores virtutis et magistri, videntur fines officiorum paulo longius quam natura vellet protulisse, ut cum ad ultimum animo contendissemus, ibi tamen, ubi oportet, consisteremus:*[50] and yet himself might have said, *Monitis sum minor ipse meis;*[51] for it was his own fault, though not in so extreme a degree.

6.   Another fault likewise much of this kind hath been incident to learned men; which is, that they have esteemed the preservation, good, and honour of their countries or masters before their own fortunes or safeties. For so saith Demosthenes unto the Athenians; *If it please you to note it, my counsels unto you are not such whereby I should grow great amongst you, and you become little amongst the Grecians: but they be of that nature, as they are sometimes not good for me to give, but are always good for you to follow.*[52] And so Seneca, after he had consecrated that *Quinquennium Neronis*[53] to the eternal glory of learned governors, held on his honest and loyal course of good and free counsel, after his master grew extremely corrupt in his government. Neither can this point otherwise be; for Learning endueth men's minds with a true sense of the frailty of their persons, the casualty of their fortunes, and the dignity of their soul and vocation: so that it is impossible for them to esteem that any greatness of their own fortune can be a true or wor-thy end of their being and ordainment; and therefore are desirous to give their account to God, and so likewise to their masters under God (as kings and the states that they serve) in these words; *Ecce tibi lu-crefeci,* and not *Ecce mihi lucrefeci;*[54] whereas the corrupter sort of mere Politiques, that have not their thoughts established by learning in the love and apprehension of duty, nor never look abroad into universal-ity, do refer all things to themselves, and thrust themselves into the centre of the world, as if all lines should meet in them and their for-tunes; never caring in all tempests what becomes of the ship of estates, so they may save themselves in the cockboat of their own fortune: whereas men that feel the weight of duty and know the limits of self-love, use to make good their places and duties, though with peril; and

if they stand in seditious and violent alterations, it is rather the reverence which many times both adverse parts do give to honesty, than any versatile advantage of their own carriage. But for this point of tender sense and fast obligation of duty which learning doth endue the mind withal, howsoever fortune may tax it, and many in the depth of their corrupt principles may despise it, yet it will receive an open allowance, and therefore needs the less disproof or excusation.

7. Another fault incident commonly to learned men, which may be more probably defended than truly denied, is, that they fail sometimes in applying themselves to particular persons: which want of exact application ariseth from two causes; the one, because the largeness of their mind can hardly confine itself to dwell in the exquisite observation or examination of the nature and customs of one person: for it is a speech for a lover, and not for a wise man: *Satis magnum alter alteri theatrum sumus.*[55] Nevertheless I shall yield, that he that cannot contract the sight of his mind as well as disperse and dilate it, wanteth a great faculty. But there is a second cause, which is no inability, but a rejection upon choice and judgment. For the honest and just bounds of observation by one person upon another, extend no further but to understand him sufficiently, whereby not to give him offence, or whereby to be able to give him faithful counsel, or whereby to stand upon reasonable guard and caution in respect of a man's self. But to be speculative into another man to the end to know how to work him, or wind him, or govern him, proceedeth from a heart that is double and cloven and not entire and ingenuous; which as in friendship it is want of integrity, so towards princes or superiors is want of duty. For the custom of the Levant, which is that subjects do forbear to gaze or fix their eyes upon princes,[56] is in the outward ceremony barbarous, but the moral is good: for men ought not by cunning and bent observations to pierce and penetrate into the hearts of kings, which the scripture hath declared to be inscrutable.[57]

8. There is yet another fault (with which I will conclude this part) which is often noted in learned men, that they do many times fail to observe decency and discretion in their behaviour and carriage, and commit errors in small and ordinary points of action, so as the vulgar sort of capacities do make a judgment of them in greater matters by

that which they find wanting in them in smaller. But this consequence doth oft deceive men, for which I do refer them over to that which was said by Themistocles, arrogantly and uncivilly being applied to himself out of his own mouth, but, being applied to the general state of this question, pertinently and justly, when, being invited to touch a lute, he said, *He could not fiddle, but he could make a small town a great state.*[58] So, no doubt, many may be well seen in the passages of government and policy, which are to seek in little and punctual occasions. I refer them also to that which Plato said of his master Socrates, whom he compared to the gallipots of apothecaries, which on the outside had apes and owls and antiques but contained within sovereign and precious liquors and confections; acknowledging that to an external report he was not without superficial levities and deformities, but was inwardly replenished with excellent virtues and powers.[59] And so much touching the point of manners of learned men.

9.   But in the mean time I have no purpose to give allowance to some conditions and courses base and unworthy, wherein divers professors of learning have wronged themselves and gone too far; such as were those trencher philosophers which in the later age of the Roman state were usually in the houses of great persons, being little better than solemn parasites; of which kind Lucian maketh a merry description of the philosopher that the great lady took to ride with her in her coach, and would needs have him carry her little dog, which he doing officiously and yet uncomely, the page scoffed and said, *That he doubted the philosopher of a Stoic would turn to be a Cynic.*[60] But above all the rest, the gross and palpable flattery, whereunto many not unlearned have abased and abused their wits and pens, turning, as Du Bartas saith,[61] Hecuba into Helena, and Faustina into Lucretia, hath most diminished the price and estimation of learning. Neither is the modern dedication of books and writings, as to patrons, to be commended: for that books, such as are worthy the name of books, ought to have no patrons but truth and reason. And the ancient custom was to dedicate them only to private and equal friends, or to entitle the books with their names: or if to kings and great persons, it was to some such as the argument of the book was fit and proper for: but these and the like courses may deserve rather reprehension than defence.

10.   Not that I can tax or condemn the morigeration or application of learned men to men in fortune. For the answer was good that Diogenes made to one that asked him in mockery, *How it came to pass that philosophers were the followers of rich men, and not rich men of philosophers?* He answered soberly, and yet sharply, *Because the one sort knew what they had need of, and the other did not.*[62] And of the like nature was the answer which Aristippus made, when having a petition to Dionysius, and no ear given to him, he fell down at his feet; whereupon Dionysius stayed and gave him the hearing, and granted it; and afterward some person, tender on the behalf of philosophy, reproved Aristippus that he would offer the profession of philosophy such an indignity as for a private suit to fall at a tyrant's feet: but he answered, *It was not his fault, but it was the fault of Dionysius that had his ears in his feet.*[63] Neither was it accounted weakness but discretion in him that would not dispute his best with Adrianus Caesar; excusing himself, *That it was reason to yield to him that commanded thirty legions.*[64] These and the like applications, and stooping to points of necessity and convenience, cannot be disallowed; for though they may have some outward baseness, yet in a judgment truly made they are to be accounted submissions to the occasion and not to the person.

IV   **1.**   Now I proceed to those errors and vanities which have intervened amongst the studies themselves of the learned, which is that which is principal and proper to the present argument; wherein my purpose is not to make a justification of the errors, but by a censure and separation of the errors to make a justification of that which is good and sound, and to deliver that from the aspersion of the other. For we see that it is the manner of men to scandalize and deprave that which retaineth the state[65] and virtue, by taking advantage upon that which is corrupt and degenerate: as the heathens in the primitive church used to blemish and taint the Christians with the faults and corruptions of heretics. But nevertheless I have no meaning at this time to make any exact animadversion of the errors and impediments in matters of learning, which are more secret and remote from vulgar opinion, but only to speak unto such as do fall under or near unto a popular observation.

2.   There be therefore chiefly three vanities in studies, whereby learning hath been most traduced. For those things we do esteem vain, which are either false or frivolous, those which either have no truth or no use: and those persons we esteem vain, which are either credulous or curious; and curiosity is either in matter or words: so that in reason, as well as in experience, there fall out to be these three distempers, as I may term them, of learning: the first, fantastical learning; the second, contentious learning; and the last, delicate learning; vain imaginations, vain altercations, and vain affectations; and with the last I will begin. Martin Luther, conducted no doubt by a higher providence, but in discourse of reason,[66] finding what a province he had undertaken against the bishop of Rome and the degenerate traditions of the church, and finding his own solitude, being no ways aided by the opinions of his own time, was enforced to awake all antiquity, and to call former times to his succours to make a party against the present time: so that the ancient authors, both in divinity and in humanity, which had long time slept in libraries, began generally to be read and revolved. This by consequence did draw on a necessity of a more exquisite travail in the languages original, wherein those authors did write, for the better understanding of those authors, and the better advantage of pressing and applying their words. And thereof grew again a delight in their manner of style and phrase, and an admiration of that kind of writing; which was much furthered and precipitated by the enmity and opposition that the propounders of those primitive but seeming new opinions had against the schoolmen; who were generally of the contrary part, and whose writings were altogether in a different style and form; taking liberty to coin and frame new terms of art to express their own sense, and to avoid circuit of speech, without regard to the pureness, pleasantness, and, as I may call it, lawfulness of the phrase or word. And again, because the great labour then was with the people (of whom the Pharisees were wont to say, *Execrabilis ista turba, quae non novit legem*)[67] for the winning and persuading of them, there grew of necessity in chief price and request eloquence and variety of discourse, as the fittest and forciblest access into the capacity of the vulgar sort:

so that these four causes concurring, the admiration of ancient authors, the hate of the schoolmen, the exact study of languages, and the efficacy of preaching, did bring in an affectionate study of eloquence and copie of speech, which then began to flourish. This grew speedily to an excess; for men began to hunt more after words than matter; more after the choiceness of the phrase, and the round and clean composition of the sentence, and the sweet falling of the clauses, and the varying and illustration of their works with tropes and figures, than after the weight of matter, worth of subject, soundness of argument, life of invention, or depth of judgment. Then grew the flowing and watery vein of Osorius[68] the Portugal bishop, to be in price. Then did Sturmius spend such infinite and curious pains upon Cicero the Orator, and Hermogenes the Rhetorician, besides his own books of Periods and Imitation, and the like. Then did Car of Cambridge and Ascham with their lectures and writings almost deify Cicero and Demosthenes, and allure all young men that were studious unto that delicate and polished kind of learning. Then did Erasmus take occasion to make the scoffing Echo: *Decem annos consumpsi in legendo Cicerone;* and the Echo answered in Greek, ὄνε; *Asine.*[69] Then grew the learning of the schoolmen to be utterly despised as barbarous. In sum, the whole inclination and bent of those times was rather towards copie than weight.

3.   Here, therefore, is the first distemper of learning, when men study words and not matter; whereof, though I have represented an example of late times, yet it hath been and will be *secundum majus et minus* in all time. And how is it possible but this should have an operation to discredit learning, even with vulgar capacities, when they see learned men's works like the first letter of a patent, or limned book; which though it hath large flourishes, yet is but a letter? It seems to me that Pygmalion's frenzy is a good emblem or portraiture of this vanity:[70] for words are but the images of matter; and except they have life of reason and invention, to fall in love with them is all one as to fall in love with a picture.

4.   But yet notwithstanding it is a thing not hastily to be condemned, to clothe and adorn the obscurity even of Philosophy itself with sensible and plausible elocution. For hereof we have great examples in Xenophon, Cicero, Seneca, Plutarch, and of Plato also in some degree;

and hereof likewise there is great use: for surely, to the severe inquisition of truth and the deep progress into philosophy, it is some hindrance; because it is too early satisfactory to the mind of man, and quencheth the desire of further search, before we come to a just period. But then if a man be to have any use of such knowledge in civil occasions, of conference, counsel, persuasion, discourse, or the like, then shall he find it prepared to his hands in those authors which write in that manner. But the excess of this is so justly contemptible that as Hercules, when he saw the image of Adonis, Venus' minion, in a temple, said in disdain, *Nil sacri es;*[71] so there is none of Hercules' followers in learning, that is, the more severe and laborious sort of inquirers into truth, but will despise those delicacies and affectations, as indeed capable of no divineness. And thus much of the first disease or distemper of learning.

5.   The second which followeth is in nature worse than the former: for as substance of matter is better than beauty of words, so contrariwise vain matter is worse than vain words: wherein it seemeth the reprehension of Saint Paul was not only proper for those times, but prophetical for the times following; and not only respective to divinity, but extensive to all knowledge; *Devita profanas vocum novitates, et oppositiones falsi nominis scientiae.*[72] For he assigneth two marks and badges of suspected and falsified science: the one, the novelty and strangeness of terms; the other, the strictness of positions, which of necessity doth induce oppositions, and so questions and altercations. Surely, like as many substances in nature which are solid do putrefy and corrupt into worms; so it is the property of good and sound knowledge to putrefy and dissolve into a number of subtle, idle, unwholesome, and, as I may term them, vermiculate questions, which have indeed a kind of quickness and life of spirit, but no soundness of matter or goodness of quality. This kind of degenerate learning did chiefly reign amongst the Schoolmen:[73] who having sharp and strong wits, and abundance of leisure, and small variety of reading, but their wits being shut up in the cells of a few authors (chiefly Aristotle their dictator) as their persons were shut up in the cells of monasteries and colleges, and knowing little history, either of nature or time, did out of no great quantity of matter and infinite agitation of wit spin out unto us those laborious webs of learning which are extant in their books.[74]

For the wit and mind of man, if it work upon matter, which is the contemplation of the creatures of God, worketh according to the stuff and is limited thereby; but if it work upon itself, as the spider worketh his web, then it is endless, and brings forth indeed cobwebs of learning, admirable for the fineness of thread and work, but of no substance or profit.

6.   This same unprofitable subtlety or curiosity is of two sorts; either in the subject itself that they handle, when it is a fruitless speculation or controversy (whereof there are no small number both in Divinity and Philosophy), or in the manner or method of handling of a knowledge, which amongst them was this; upon every particular position or assertion to frame objections, and to those objections, solutions; which solutions were for the most part not confutations but distinctions: whereas indeed the strength of all sciences is, as the strength of the old man's fagot, in the bond. For the harmony of a science, supporting each part the other, is and ought to be the true and brief confutation and suppression of all the smaller sort of objections. But, on the other side, if you take out every axiom, as the sticks of the fagot, one by one, you may quarrel with them and bend them and break them at your pleasure: so that, as was said of Seneca, *Verborum minutiis rerum frangit pondera;*[75] so a man may truly say of the schoolmen, *Quaestionum minutiis scientiarum frangunt soliditatem.* For were it not better for a man in a fair room to set up one great light, or branching candlestick of lights, than to go about with a small watch candle into every corner? And such is their method, that rests not so much upon evidence of truth proved by arguments, authorities, similitudes, examples, as upon particular confutations and solutions of every scruple, cavillation, and objection; breeding for the most part one question as fast as it solveth another; even as in the former resemblance, when you carry the light into one corner, you darken the rest; so that the fable and fiction of Scylla seemeth to be a lively image of this kind of philosophy or knowledge; which was transformed into a comely virgin for the upper parts; but then *Candida succinctam latranibus inguina monstris:*[76] so the generalities of the schoolmen are for a while good and proportionable; but then, when you descend into their distinctions and decisions, instead of a fruitful womb for the use and benefit of

man's life, they end in monstrous altercations and barking questions. So as it is not possible but this quality of knowledge must fall under popular contempt, the people being apt to contemn truth upon occasion of controversies and altercations, and to think they are all out of their way which never meet; and when they see such digladiation about subtilties, and matters of no use or moment, they easily fall upon that judgment of Dionysius of Syracuse, *Verba ista sunt senum otiosorum.*[77]

7. Notwithstanding, certain it is that if those Schoolmen to their great thirst of truth and unwearied travail of wit had joined variety and universality of reading and contemplation, they had proved excellent lights, to the great advancement of all learning and knowledge: but as they are, they are great undertakers indeed, and fierce with dark keeping. But as in the inquiry of the divine truth, their pride inclined to leave the oracle of God's word, and to vanish in the mixture of their own inventions; so in the inquisition of nature, they ever left the oracle of God's works, and adored the deceiving and deformed images which the unequal mirror of their own minds, or a few received authors or principles, did represent unto them. And thus much for the second disease of learning.

8. For the third vice or disease of learning, which concerneth deceit or untruth, it is of all the rest the foulest; as that which doth destroy the essential form of knowledge, which is nothing but a representation of truth: for the truth of being and the truth of knowing are one, differing no more than the direct beam and the beam reflected. This vice therefore brancheth itself into two sorts; delight in deceiving and aptness to be deceived; imposture and credulity; which, although they appear to be of a diverse nature, the one seeming to proceed of cunning and the other of simplicity, yet certainly they do for the most part concur: for, as the verse noteth,

Percontatorem fugito, nam garrulus idem est:[78]

an inquisitive man is a prattler; so upon the like reason a credulous man is a deceiver: as we see it in fame, that he that will easily believe rumours, will as easily augment rumours, and add somewhat to them of his own; which Tacitus wisely noteth, when he saith, *Fingunt simul creduntque:*[79] so great an affinity hath fiction and belief.

9.   This facility of credit and accepting or admitting things weakly authorised or warranted, is of two kinds according to the subject: for it is either a belief of history, or, as the lawyers speak, matter of fact; or else of matter of art and opinion. As to the former, we see the experience and inconvenience of this error in ecclesiastical history; which hath too easily received and registered reports and narrations of miracles wrought by martyrs, hermits, or monks of the desert, and other holy men, and their relics, shrines, chapels, and images: which though they had a passage for a time by the ignorance of the people, the superstitious simplicity of some, and the politic toleration of others holding them but as divine poesies; yet after a period of time, when the mist began to clear up, they grew to be esteemed but as old wives' fables, impostures of the clergy, illusions of spirits, and badges of Antichrist, to the great scandal and detriment of religion.

10.   So in natural history, we see there hath not been that choice and judgment used as ought to have been; as may appear in the writings of Plinius, Cardanus,[80] Albertus,[81] and divers of the Arabians, being fraught with much fabulous matter, a great part not only untried, but notoriously untrue, to the great derogation of the credit of natural philosophy with the grave and sober kind of wits: wherein the wisdom and integrity of Aristotle is worthy to be observed; that, having made so diligent and exquisite a history of living creatures, hath mingled it sparingly with any vain or feigned matter: and yet on the other side, hath cast all prodigious narrations, which he thought worthy the recording, into one book:[82] excellently discerning that matter of manifest truth (such whereupon observation and rule were to be built), was not to be mingled or weakened with matter of doubtful credit; and yet again, that rarities and reports that seem incredible are not to be suppressed or denied to the memory of men.

11.   And as for the facility of credit which is yielded to arts and opinions, it is likewise of two kinds; either when too much belief is attributed to the arts themselves, or to certain authors in any art. The sciences themselves, which have had better intelligence and confederacy with the imagination of man than with his reason, are three in number; astrology, natural magic, and alchemy: of which sciences, nevertheless, the ends or pretences are noble. For astrology pretendeth to discover that correspondence or concatenation which is between the superior globe and the inferior: natural magic pretendeth to call and reduce natural philosophy from variety of speculations to the magnitude of works: and alchemy pretendeth to make separation of all the unlike parts of bodies which in mixtures of nature are incorporate. But the derivations and prosecutions to these ends, both in the theories and in the practices, are full of error and vanity; which the great professors themselves have sought to veil over and conceal by enigmatical writings, and referring themselves to auricular traditions and such other devices, to save the credit of impostures: and yet surely to alchemy this right is due, that it may be compared to the husbandman whereof Aesop makes the fable; that, when he died, told his sons that he had left unto them gold buried under ground in his vineyard; and they digged over all the ground, and gold they found none; but by reason of their stirring and digging the mould about the roots of their vines, they had a great vintage the year following: so assuredly the search and stir to make gold hath brought to light a great number of good and fruitful inventions and experiments, as well for the disclosing of nature as for the use of man's life.

12. And as for the overmuch credit that hath been given unto authors in sciences, in making them dictators, that their words should stand, and not consuls to give advice; the damage is infinite that sciences have received thereby, as the principal cause that hath kept them low at a stay without growth or advancement. For hence it hath come, that in arts mechanical the first deviser comes shortest, and time addeth and perfecteth; but in sciences the first author goeth furthest, and time leeseth and corrupteth. So we see, artillery, sailing, printing, and the like, were grossly managed at the first, and by time accommodated and refined: but contrariwise, the philosophies and sciences of Aristotle, Plato, Democritus, Hippocrates, Euclides, Archimedes, of most vigour at the first and by time degenerate and imbased; whereof the reason is no other, but that in the former many wits and industries have contributed in one; and in the latter many wits and industries have been spent about the wit of some one, whom many times they have rather depraved than illustrated. For as water will not ascend higher than the level of the first springhead from whence it descendeth, so knowledge derived from Aristotle, and exempted from liberty of examination, will not rise again higher than the knowledge of Aristotle. And therefore although the position be good, *Oportet discentem credere,*[83] yet it must be coupled with this, *Oportet edoctum judicare;* for disciples do owe unto masters only a temporary belief and a suspension of their own judgment till they be fully instructed, and not an absolute resignation or perpetual captivity: and therefore, to conclude this point, I will say no more, but so let great authors have their due, as time, which is the author of authors, be not deprived of his due, which is, further and further to discover truth. Thus have I gone over these three diseases of learning; besides the which there are some other rather peccant humours than formed diseases, which nevertheless are not so secret and intrinsic but that they fall under a popular observation and traducement, and therefore are not to be passed over.

V 1.   The first of these is the extreme affecting of two extremities: the one antiquity, the other novelty; wherein it seemeth the children of time do take after the nature and malice of the father. For as he devoureth his children, so one of them seeketh to devour and suppress the other; while antiquity envieth there should be new additions, and novelty cannot be content to add but it must deface. Surely the advice of the prophet is the true direction in this matter, *State super vias antiquas, et videte quaenam sit via recta et bona et ambulate in ea.*[84] Antiquity deserveth that reverence, that men should make a stand thereupon and discover what is the best way; but when the discovery is well taken, then to make progression. And to speak truly, *Antiquitas saeculi juventus mundi.*[85] These times are the ancient times, when the world is ancient, and not those which we account ancient *ordine retrogrado,* by a computation backward from ourselves.

2.   Another error induced by the former is a distrust that anything should be now to be found out, which the world should have missed and passed over so long time; as if the same objection were to be made to time, that Lucian maketh to Jupiter and other the heathen gods; of which he wondereth that they begot so many children in old time, and begot none in his time; and asketh whether they were become septua-

genary, or whether the law *Papia,* made against old men's marriages, had restrained them. So it seemeth men doubt lest time is become past children and generation; wherein contrariwise we see commonly the levity and inconstancy of men's judgments, which till a matter be done, wonder that it can be done; and as soon as it is done, wonder again that it was no sooner done: as we see in the expedition of Alexander into Asia, which at first was prejudged as a vast and impossible enterprise; and yet afterwards it pleaseth Livy to make no more of it than this: *Nil aliud quam bene ausus vana contemnere.*[86] And the same happened to Columbus in the western navigation. But in intellectual matters it is much more common; as may be seen in most of the propositions of Euclid; which till they be demonstrate, they seem strange to our assent; but being demonstrate, our mind accepteth of them by a kind of relation (as the lawyers speak), as if we had known them before.

3. Another error, that hath also some affinity with the former, is a conceit that of former opinions or sects, after variety and examination, the best hath still prevailed and suppressed the rest; so as if a man should begin the labour of a new search, he were but like to light upon somewhat formerly rejected, and by rejection brought into oblivion: as if the multitude, or the wisest for the multitude's sake, were not ready to give passage rather to that which is popular and superficial than to that which is substantial and profound; for the truth is that time seemeth to be of the nature of a river or stream, which carrieth down to us that which is light and blown up, and sinketh and drowneth that which is weighty and solid.

4. Another error, of a diverse nature from all the former, is the over-early and peremptory reduction of knowledge into arts and methods; from which time commonly sciences receive small or no augmentation. But as young men, when they knit and shape perfectly, do seldom grow to a further stature; so knowledge, while it is in aphorisms and observations, it is in growth: but when it once is comprehended in exact methods, it may perchance be further polished and illustrate and accommodated for use and practice; but it increaseth no more in bulk and substance.

5. Another error, which doth succeed that which we last mentioned,

is that after the distribution of particular arts and sciences, men have abandoned universality, or *philosophia prima;* which cannot but cease and stop all progression. For no perfect discovery can be made upon a flat or a level: neither is it possible to discover the more remote and deeper parts of any science, if you stand but upon the level of the same science and ascend not to a higher science.

6.　Another error hath proceeded from too great a reverence, and a kind of adoration of the mind and understanding of man; by means whereof men have withdrawn themselves too much from the contemplation of nature, and the observations of experience, and have tumbled up and down in their own reason and conceits. Upon these intellectualists, which are notwithstanding commonly taken for the most sublime and divine philosophers, Heraclitus gave a just censure, saying, *Men sought truth in their own little worlds, and not in the great and common world;*[87] for they disdain to spell, and so by degrees to read in the volume of God's works: and contrariwise by continual meditation and agitation of wit do urge and as it were invocate their own spirits to divine and give oracles unto them, whereby they are deservedly deluded.

7.　Another error that hath some connection with this latter is that men have used to infect their meditations, opinions, and doctrines, with some conceits which they have most admired, or some sciences which they have most applied; and given all things else a tincture according to them, utterly untrue and unproper. So hath Plato intermingled his philosophy with theology, and Aristotle with logic; and the second school of Plato, Proclus and the rest, with the mathematics.[88] For these were the arts which had a kind of primogeniture with them severally. So have the alchemists made a philosophy out of a few experiments of the furnace; and Gilbertus,[89] our countryman, hath made a philosophy out of the observations of a lodestone. So Cicero, when, reciting the several opinions of the nature of the soul, he found a musician that held the soul was but a harmony, saith pleasantly, *Hic ab arte sua non recessit,* etc.[90] But of these conceits Aristotle speaketh seriously and wisely when he saith, *Qui respiciunt ad pauca de facili pronunciant.*[91]

8.　Another error is an impatience of doubt, and haste to assertion without due and mature suspension of judgment. For the two ways of contemplation are not unlike the two ways of action commonly spoken of by the ancients: the one plain and smooth in the beginning, and in the end impassable; the other rough and troublesome in the entrance, but after a while fair and even. So it is in contemplation; if a man will begin with certainties, he shall end in doubts; but if he will be content to begin with doubts, he shall end in certainties.

9.   Another error is in the manner of the tradition and delivery of knowledge, which is for the most part magistral and peremptory, and not ingenuous and faithful; in a sort as may be soonest believed, and not easiliest examined. It is true that in compendious treatises for practice that form is not to be disallowed: but in the true handling of knowledge, men ought not to fall either on the one side into the vein of Velleius the Epicurean: *Nil tam metuens, quam ne dubitare aliqua de re videretur;*[92] nor on the other side into Socrates his ironical doubting of all things;[93] but to propound things sincerely with more or less asseveration, as they stand in a man's own judgment proved more or less.

10.   Other errors there are in the scope that men propound to themselves, whereunto they bend their endeavours; for whereas the more constant and devote kind of professors of any science ought to propound to themselves to make some additions to their science, they convert their labours to aspire to certain second prizes: as to be a profound interpreter or commenter, to be a sharp champion or defender, to be a methodical compounder or abridger; and so the patrimony of knowledge cometh to be sometimes improved, but seldom augmented.

11.   But the greatest error of all the rest is the mistaking or misplacing of the last or furthest end of knowledge: for men have entered into a desire of learning and knowledge, sometimes upon a natural curiosity and inquisitive appetite; sometimes to entertain their minds with variety and delight; sometimes for ornament and reputation; and sometimes to enable them to victory of wit and contradiction; and most times for lucre and profession; and seldom sincerely to give a true account of their gift of reason, to the benefit and use of men: as if there were sought in knowledge a couch whereupon to rest a searching and restless spirit; or a terrace for a wandering and variable mind to walk up and down with a fair prospect; or a tower of state for a proud mind

to raise itself upon; or a fort or commanding ground for strife and contention; or a shop for profit or sale; and not a rich storehouse for the glory of the Creator and the relief of man's estate. But this is that which will indeed dignify and exalt knowledge, if contemplation and action may be more nearly and straitly conjoined and united together than they have been; a conjunction like unto that of the two highest planets, Saturn, the planet of rest and contemplation, and Jupiter, the planet of civil society and action: howbeit, I do not mean, when I speak of use and action, that end before-mentioned of the applying of knowledge to lucre and profession; for I am not ignorant how much that diverteth and interrupteth the prosecution and advancement of knowledge, like unto the golden ball thrown before Atalanta, which while she goeth aside and stoopeth to take up, the race is hindered;

Declinat cursus, aurumque volubile tollit.[94]

Neither is my meaning, as was spoken of Socrates, to call philosophy down from heaven to converse upon the earth;[95] that is, to leave natural philosophy aside, and to apply knowledge only to manners and policy. But as both heaven and earth do conspire and contribute to the use and benefit of man; so the end ought to be, from both philosophies to separate and reject vain speculations, and whatsoever is empty and void, and to preserve and augment whatsoever is solid and fruitful: that knowledge may not be as a courtesan, for pleasure and vanity only, or as a bond-woman, to acquire and gain to her master's use; but as a spouse, for generation, fruit, and comfort.

12.   Thus have I described and opened, as by a kind of dissection, those peccant humours (the principal of them) which hath[96] not only given impediment to the proficience of learning, but have given also occasion to the traducement thereof: wherein if I have been too plain, it must be remembered, *fidelia vulnera amantis, sed dolosa oscula malignantis.*[97] This I think I have gained, that I ought to be the better believed in that which I shall say pertaining to commendation; because I have proceeded so freely in that which concerneth censure. And yet I have no purpose to enter into a laudative of learning, or to make a hymn to the Muses (though I am of opinion that it is long since their rites were duly celebrated), but my intent is, without varnish or ampli-

fication justly to weigh the dignity of knowledge in the balance with other things, and to take the true value thereof by testimonies and arguments divine and human.

VI **1.** First therefore let us seek the dignity of knowledge in the archetype or first platform, which is in the attributes and acts of God, as far as they are revealed to man and may be observed with sobriety; wherein we may not seek it by the name of Learning; for all Learning is Knowledge acquired, and all knowledge in God is original: and therefore we must look for it by another name, that of Wisdom or Sapience, as the Scriptures call it.

2. It is so then, that in the work of the creation we see a double emanation of Virtue from God; the one referring more properly to Power, the other to Wisdom; the one expressed in making the subsistence of the matter, and the other in disposing the beauty of the form. This being supposed, it is to be observed that for anything which appeareth in the history of the creation, the confused mass and matter of Heaven and Earth was made in a moment; and the order and disposition of that chaos or mass was the work of six days; such a note of difference it pleased God to put upon the works of Power, and the works of Wisdom; wherewith concurreth, that in the former it is not set down that God said, *Let there be heaven and earth,* as it is set down of the works following; but actually, that God made Heaven and Earth: the one carrying the style of a Manufacture, and the other of a Law, Decree, or Counsel.

3. To proceed to that which is next in order from God to Spirits:[98] we find, as far as credit is to be given to the celestial hierarchy of that supposed Dionysius the senator of Athens, the first place or degree is given to the angels of Love, which are termed Seraphim; the second to the angels of Light, which are termed Cherubim; and the third, and so following places, to Thrones, Principalities, and the rest, which are all angels of power and ministry; so as the angels of Knowledge and Illumination are placed before the angels of Office and Domination.[99]

4. To descend from Spirits and Intellectual Forms to Sensible and Material Forms, we read the first Form that was created was Light,[100] which hath a relation and correspondence in nature and corporal things to Knowledge in Spirits and incorporal things.

5. So in the distribution of days we see the day wherein God did rest and contemplate His own works, was blessed above all the days wherein He did effect and accomplish them.[101]

6. After the creation was finished, it is set down unto us that man was placed in the garden to work therein; which work, so appointed to him, could be no other than work of Contemplation; that is, when the end of work is but for exercise and experiment, not for necessity; for there being then no reluctation of the creature, nor sweat of the brow, man's employment must of consequence have been matter of delight in the experiment, and not matter of labour for the use. Again, the first acts which man performed in Paradise consisted of the two summary parts of knowledge; the view of creatures, and the imposition of names.[102] As for the knowledge which induced the fall, it was, as was touched before, not the natural knowledge of creatures, but the moral knowledge of good and evil; wherein the supposition was that God's commandments or prohibitions were not the originals of good and evil, but that they had other beginnings, which man aspired to know; to the end to make a total defection from God and to depend wholly upon himself.

7. To pass on: in the first event or occurrence after the fall of man, we see (as the Scriptures have infinite mysteries, not violating at all the truth of the story or letter) an image of the two estates, the contemplative state and the active state, figured in the two persons of Abel and Cain, and in the two simplest and most primitive trades of life; that of the shepherd (who, by reason of his leisure, rest in a place, and living in view of heaven, is a lively image of a contemplative life), and that of the husbandman:[103] where we see again the favour and election of God went to the shepherd, and not to the tiller of the ground.

8. So in the age before the flood, the holy records within those few memorials which are there entered and registered have vouchsafed to mention and honour the name of the inventors and authors of music and works in metal.[104] In the age after the flood, the first great judgment of God upon the ambition of man was the confusion of tongues;[105]

whereby the open trade and intercourse of learning and knowledge was chiefly imbarred.

9.   To descend to Moses the lawgiver, and God's first pen: he is adorned by the Scriptures with this addition and commendation, *That he was seen in all the learning of the Egyptians;*[106] which nation we know was one of the most ancient schools of the world: for so Plato brings in the Egyptian priest saying unto Solon: *You Grecians are ever children; you have no knowledge of antiquity, nor antiquity of knowledge.*[107] Take a view of the ceremonial law of Moses; you shall find, besides the prefiguration of Christ, the badge or difference of the people of God, the exercise and impression of obedience, and other divine uses and fruits thereof, that some of the most learned Rabbins have travailed profitably and profoundly to observe, some of them a natural, some of them a moral sense or reduction of many of the ceremonies and ordinances. As in the law of the leprosy, where it is said, *If the whiteness have overspread the flesh, the patient may pass abroad for clean; but if there be any whole flesh remaining, he is to be shut up for unclean;*[108] one of them noteth a principle of nature, that *putrefaction is more contagious before maturity than after:* and another noteth a position of moral philosophy, that *men abandoned to vice do not so much corrupt manners, as those that are half good and half evil.* So in this and very many other places in that law, there is to be found, besides the theological sense, much aspersion of philosophy.

10.   So likewise in that excellent book of Job, if it be revolved with diligence, it will be found pregnant and swelling with natural philosophy; as, for example, cosmography, and the roundness of the world, *Qui extendit aquilonem super vacuum, et appendit terram super nihilum;*[109] wherein the pensileness of the earth, the pole of the north, and the finiteness or convexity of heaven are manifestly touched. So again, matter of astronomy; *Spiritus ejus ornavit caelos, et obstetricante manu ejus eductus est coluber tortuosus.*[110] And in another place, *Nunquid conjungere valebis micantes stellas Pleiadas, aut gyrum Arcturi poteris dissipare?*[111] Where the fixing of the stars, ever standing at equal distance, is with great elegancy noted. And in another place, *Qui facit Arcturum, et Oriona, et Hyadas, et interiora Austri;*[112] where again he takes knowledge of the depression of the southern pole, calling it the secrets

of the south, because the southern stars were in that climate unseen. Matter of generation; *Annon sicut lac mulsisti me, et sicut caseum coagulasti me?* etc.[113] Matter of minerals; *Habet argentum venarum suarum principia: et auro locus est in quo conflatur, ferrum de terra tollitur, et lapis solutus calore in aes vertitur:*[114] and so forwards in that chapter.

11. So likewise in the person of Salomon the King, we see the gift or endowment of wisdom and learning, both in Salomon's petition and in God's assent thereunto, preferred before all other terrene and temporal felicity.[115] By virtue of which grant or donative of God Salomon became enabled not only to write those excellent Parables or Aphorisms concerning divine and moral philosophy; but also to compile a Natural History of all verdure, from the cedar upon the mountain to the moss upon the wall (which is but a rudiment between putrefaction and a herb),[116] and also of all things that breathe or move.[117] Nay, the same Salomon the King, although he excelled in the glory of treasure and magnificent buildings, of shipping and navigation, of service and attendance, of fame and renown, and the like, yet he maketh no claim to any of those glories, but only to the glory of inquisition of truth; for so he saith expressly, *The glory of God is to conceal a thing, but the glory of the king is to find it out;*[118] as if, according to the innocent play of children, the Divine Majesty took delight to hide His works, to the end to have them found out; and as if kings could not obtain a greater honour than to be God's playfellows in that game; considering the great commandment of wits and means, whereby nothing needeth to be hidden from them.

12. Neither did the dispensation of God vary in the times after our Saviour came into the world; for our Saviour Himself did first show His power to subdue ignorance, by His conference with the priests and doctors of the law,[119] before He showed His power to subdue nature by His miracles. And the coming of the Holy Spirit was chiefly figured and expressed in the similitude and gift of tongues,[120] which are but *vehicula scientiae.*

13. So in the election of those instruments, which it pleased God to use for the plantation of the Faith, notwithstanding that at the first He did employ persons altogether unlearned, otherwise than by inspiration, more evidently to declare His immediate working, and to abase

all human wisdom or knowledge; yet nevertheless that counsel of His was no sooner performed, but in the next vicissitude and succession He did send His Divine Truth into the world, waited on with other learnings, as with servants or handmaids: for so we see Saint Paul, who was the only learned amongst the Apostles, had his pen most used in the Scriptures of the New Testament.

14.   So again we find that many of the ancient Bishops and Fathers of the Church were excellently read and studied in all the learning of the heathen; insomuch that the edict of the Emperor Julianus,[121] whereby it was interdicted unto Christians to be admitted into schools, lectures, or exercises of learning, was esteemed and accounted a more pernicious engine and machination against the Christian Faith, than were all the sanguinary prosecutions of his predecessors; neither could the emulation and jealousy of Gregory the first of that name, bishop of Rome,[122] ever obtain the opinion of piety or devotion; but contrariwise received the censure of humour, malignity, and pusillanimity, even amongst holy men; in that he designed to obliterate and extinguish the memory of heathen antiquity and authors. But contrariwise it was the Christian Church, which, amidst the inundations of the Scythians on the one side from the northwest and the Saracens from the east, did preserve in the sacred lap and bosom thereof the precious relics even of heathen learning, which otherwise had been extinguished as if no such thing had ever been.

15.   And we see before our eyes, that in the age of ourselves and our fathers, when it pleased God to call the Church of Rome to account for their degenerate manners and ceremonies, and sundry doctrines obnoxious and framed to uphold the same abuses; at one and the same time it was ordained by the Divine Providence that there should attend withal a renovation and new spring of all other knowledges. And on the other side we see the Jesuits (who partly in themselves, and partly by the emulation and provocation of their example, have much quickened and strengthened the state of learning), we see, I say, what notable service and reparation they have done to the Roman see.

16.   Wherefore to conclude this part, let it be observed, that there be two principal duties and services, besides ornament and illustration, which philosophy and human learning do perform to faith and

religion. The one, because they are an effectual inducement to the exaltation of the glory of God: for as the Psalms and other Scriptures do often invite us to consider and magnify the great and wonderful works of God,[123] so if we should rest only in the contemplation of the exterior of them as they first offer themselves to our senses, we should do a like injury unto the Majesty of God, as if we should judge or construe of the store of some excellent jeweller, by that only which is set out toward the street in his shop. The other, because they minister a singular help and preservative against unbelief and error: for our Saviour saith, *You err, not knowing the Scriptures, nor the power of God;*[124] laying before us two books or volumes to study, if we will be secured from error; first the Scriptures, revealing the Will of God; and then the creatures expressing His Power;[125] whereof the latter is a key unto the former: not only opening our understanding to conceive the true sense of the Scriptures, by the general notions of reason and rules of speech; but chiefly opening our belief, in drawing us into a due meditation of the omnipotency of God, which is chiefly signed and engraven upon His works. Thus much therefore for divine testimony and evidence concerning the true dignity and value of Learning.

VII

1.   As for human proofs, it is so large a field, as in a discourse of this nature and brevity it is fit rather to use choice of those things which we shall produce, than to embrace the variety of them. First therefore, in the degrees of human honour amongst the heathen, it was the highest to obtain to a veneration and adoration as a God. This unto the Christians is as the forbidden fruit. But we speak now separately of human testimony: according to which, that which the Grecians call *apotheosis,* and the Latins *relatio inter divos,* was the supreme honour which man could attribute unto man: specially when it was given, not by a formal decree or act of state, as it was used among the Roman Emperors, but by an inward assent and belief. Which honour, being so high, had also a degree or middle term; for there were reckoned above human honours, honours heroical and divine: in the attribution and distribution of which honours we see antiquity made this difference: that whereas founders and uniters of states and cities,

law-givers, extirpers of tyrants, fathers of the people, and other eminent persons in civil merit, were honoured but with the titles of worthies or demi-gods; such as were Hercules, Theseus, Minos, Romulus, and the like: on the other side, such as were inventors and authors of new arts, endowments, and commodities towards man's life, were ever consecrated amongst the gods themselves; as were Ceres, Bacchus, Mercurius, Apollo, and others: and justly; for the merit of the former is confined within the circle of an age or a nation; and is like fruitful showers, which though they be profitable and good, yet serve but for that season, and for a latitude of ground where they fall; but the other is indeed like the benefits of heaven, which are permanent and universal. The former again is mixed with strife and perturbation; but the latter hath the true character of Divine Presence, coming in *aura leni*, without noise or agitation.

2.   Neither is certainly that other merit of learning, in repressing the inconveniences which grow from man to man, much inferior to the former, of relieving the necessities which arise from nature; which merit was lively set forth by the ancients in that feigned relation of Orpheus' theatre, where all beasts and birds assembled; and forgetting their several appetites, some of prey, some of game, some of quarrel, stood all sociably together listening unto the airs and accords of the harp; the sound whereof no sooner ceased, or was drowned by some louder noise, but every beast returned to his own nature: wherein is aptly described the nature and condition of men, who are full of savage and unreclaimed desires, of profit, of lust, of revenge; which as long as they give ear to precepts, to laws, to religion, sweetly touched with eloquence and persuasion of books, of sermons, of harangues, so long is society and peace maintained; but if these instruments be silent, or that sedition and tumult make them not audible, all things dissolve into anarchy and confusion.

3.   But this appeareth more manifestly, when kings themselves, or persons of authority under them, or other governors in commonwealths and popular estates, are endued with learning. For although he might be thought partial to his own profession, that said, *Then should people and estates be happy, when either kings were philosophers, or philosophers kings;*[126] yet so much is verified by experience, that under

learned princes and governors there have been ever the best times: for howsoever kings may have their imperfections in their passions and customs; yet if they be illuminate by learning, they have those notions of religion, policy, and morality, which do preserve them and refrain them from all ruinous and peremptory errors and excesses; whispering evermore in their ears, when counsellors and servants stand mute and silent. And senators or counsellors likewise, which be learned, do proceed upon more safe and substantial principles, than counsellors which are only men of experience: the one sort keeping dangers afar off, whereas the other discover them not till they come near hand, and then trust to the agility of their wit to ward or avoid them.

4.　Which felicity of times under learned princes (to keep still the law of brevity, by using the most eminent and selected examples) doth best appear in the age which passed from the death of Domitian the emperor until the reign of Commodus; comprehending a succession of six princes, all learned, or singular favourers and advancers of learning, which age for temporal respects was the most happy and flourishing that ever the Roman empire (which then was a model of the world) enjoyed: a matter revealed and prefigured unto Domitian in a dream the night before he was slain; for he thought there was grown behind upon his shoulders a neck and a head of gold: which came accordingly to pass in those golden times which succeeded: of which princes we will make some commemoration; wherein although the matter will be vulgar, and may be thought fitter for a declamation than agreeable to a treatise infolded as this is, yet because it is pertinent to the point in hand, *Neque semper arcum tendit Apollo,*[127] and to name them only were too naked and cursory, I will not omit it altogether. The first was Nerva; the excellent temper of whose government is by a glance in Cornelius Tacitus touched to the life: *Postquam divus Nerva res olim insociabiles miscuisset, imperium et libertatem.*[128] And in token of his learning, the last act of his short reign left to memory was a missive to his adopted son Trajan, proceeding upon some inward discontent at the ingratitude of the times, comprehended in a verse of Homer's:

Telis, Phoebe, tuis lacrymas ulciscere nostras.[129]

5.　Trajan, who succeeded, was for his person not learned: but if we will hearken to the speech of our Saviour, that saith, *He that receiveth a prophet in the name of a prophet shall have a prophet's reward,*[130] he deserveth to be placed amongst the most learned princes: for there was not a greater admirer of learning or benefactor of learning; a founder of famous libraries, a perpetual advancer of learned men to office, and a familiar converser with learned professors and preceptors, who were noted to have then most credit in court. On the other side, how much Trajan's virtue and government was admired and renowned, surely no testimony of grave and faithful history doth more lively set forth than that legend tale of Gregorius Magnus, bishop of Rome, who was noted for the extreme envy he bore towards all heathen excellency: and yet he is reported, out of the love and estimation of Trajan's moral virtues, to have made unto God passionate and fervent prayers for the delivery of his soul out of hell:[131] and to have obtained it, with a caveat that he should make no more such petitions. In this prince's time also the persecutions against the Christians received intermission, upon the certificate of Plinius Secundus, a man of excellent learning and by Trajan advanced.[132]

6.   Adrian, his successor, was the most curious man that lived, and the most universal inquirer; insomuch as it was noted for an error in his mind, that he desired to comprehend all things, and not to reserve himself for the worthiest thing: falling into the like humour that was long before noted in Philip of Macedon, who, when he would needs over-rule and put down an excellent musician in an argument touching music, was well answered by him again, *God forbid, sir,* saith he, *that your fortune should be so bad, as to know these things better than I.*[133] It pleased God likewise to use the curiosity of this emperor as an inducement to the peace of His Church in those days. For having Christ in veneration, not as a God or Saviour, but as a wonder or novelty; and having His picture in his gallery, matched with Apollonius, with whom in his vain imagination he thought he had some conformity; yet it served the turn to allay the bitter hatred of those times against the Christian name, so as the Church had peace during his time. And for his government civil, although he did not attain to that of Trajan's in glory of arms or perfection of justice, yet in deserving of the weal of the subject he did exceed him. For Trajan erected many famous monu-

ments and buildings; insomuch as Constantine the Great in emulation was wont to call him Parietaria, wall-flower, because his name was upon so many walls: but his buildings and works were more of glory and triumph than use and necessity. But Adrian spent his whole reign, which was peaceable, in a perambulation or survey of the Roman empire; giving order and making assignation where he went, for re-edifying of cities, towns, and forts decayed; and for cutting of rivers and streams, and for making bridges and passages, and for policing of cities and commonalties with new ordinances and constitutions, and granting new franchises and incorporations; so that his whole time was a very restoration of all the lapses and decays of former times.

7.   Antoninus Pius, who succeeded him, was a prince excellently learned, and had the patient and subtle wit of a schoolman; insomuch as in common speech, which leaves no virtue untaxed, he was called Cymini Sector,[134] a carver or divider of cumin seed, which is one of the least seeds; such a patience he had and settled spirit to enter into the least and most exact differences of causes; a fruit no doubt of the exceeding tranquillity and serenity of his mind; which being no ways charged or incumbered, either with fears, remorses, or scruples, but having been noted for a man of the purest goodness, without all fiction or affectation, that hath reigned or lived, made his mind continually present and entire. He likewise approached a degree nearer unto Christianity, and became, as Agrippa said unto Saint Paul, half a Christian;[135] holding their religion and law in good opinion, and not only ceasing persecution, but giving way to the advancement of Christians.

8.   There succeeded him the first *Divi fratres,* the two adoptive brethren, Lucius Commodus Verus[136] (son to Aelius Verus, who delighted much in the softer kind of learning, and was wont to call the poet Martial his Virgil[137]) and Marcus Aurelius Antoninus; whereof the latter, who obscured his colleague and survived him long, was named the Philosopher: who, as he excelled all the rest in learning, so he excelled them likewise in perfection of all royal virtues; insomuch as Julianus the emperor, in his book entitled *Caesares,* being as a pasquil or satire to deride all his predecessors, feigned that they were all invited to a banquet of the gods, and Silenus the jester sat at the nether end of the table, and bestowed a scoff on every one as they came in; but

when Marcus Philosophus came in, Silenus was gravelled and out of countenance, not knowing where to carp at him; save at the last he gave a glance at his patience towards his wife. And the virtue of this prince, continued with that of his predecessor, made the name of Antoninus so sacred in the world, that though it were extremely dishonoured in Commodus, Caracalla, and Heliogabalus, who all bore the name, yet when Alexander Severus refused the name because he was a stranger to the family, the senate with one acclamation said, *Quomodo Augustus, sic et Antoninus.* In such renown and veneration was the name of these two princes in those days, that they would have it as a perpetual addition in all the emperors' style. In this emperor's time also the Church for the most part was in peace; so as in this sequence of six princes we do see the blessed effects of learning in sovereignty, painted forth in the greatest table of the world.

9. But for a tablet or picture of smaller volume (not presuming to speak of your Majesty that liveth), in my judgment the most excellent is that of Queen Elizabeth, your immediate predecessor in this part of Britain; a prince that, if Plutarch were now alive to write lives by parallels, would trouble him, I think, to find for her a parallel amongst women. This lady was endued with learning in her sex singular, and rare even amongst masculine princes; whether we speak of learning, of language, or of science, modern or ancient, Divinity or Humanity: and unto the very last year of her life she was accustomed to appoint set hours for reading, scarcely any young student in a university more daily, or more duly. As for her government, I assure myself I shall not exceed, if I do affirm that this part of the island never had forty-five years of better times; and yet not through the calmness of the season, but through the wisdom of her regiment. For if there be considered of the one side, the truth of religion established; the constant peace and security; the good administration of justice; the temperate use of the prerogative, not slackened, nor much strained; the flourishing state of learning, sortable to so excellent a patroness; the convenient estate of wealth and means, both of Crown and subject; the habit of obedience, and the moderation of discontents: and there be considered on the other side the differences of religion; the troubles of neighbour countries; the ambition of Spain, and opposition of Rome; and then, that

she was solitary and of herself: these things I say considered, as I could not have chosen an instance so recent and so proper, so I suppose I could not have chosen one more remarkable or eminent to the purpose now in hand, which is concerning the conjunction of learning in the prince with felicity in the people.

10.   Neither hath learning an influence and operation only upon civil merit and moral virtue, and the arts or temperature of peace and peaceable government; but likewise it hath no less power and efficacy in enablement towards martial and military virtue and prowess; as may be notably represented in the examples of Alexander the Great and Caesar the dictator, mentioned before, but now in fit place to be resumed: of whose virtues and acts in war there needs no note or recital, having been the wonders of time in that kind: but of their affections towards learning, and perfections in learning, it is pertinent to say somewhat.

11.   Alexander[138] was bred and taught under Aristotle, the great philosopher, who dedicated divers of his books of philosophy unto him: he was attended with Callisthenes and divers other learned persons, that followed him in camp, throughout his journeys and conquests. What price and estimation he had learning in doth notably appear in these three particulars: first, in the envy he used to express that he bore towards Achilles, in this, that he had so good a trumpet of his praises as Homer's verses; secondly, in the judgment or solution he gave touching that precious cabinet of Darius, which was found among his jewels; whereof question was made what thing was worthy to be put into it; and he gave his opinion for Homer's works: thirdly, in his letter to Aristotle, after he had set forth his books of nature, wherein he expostulateth with him for publishing the secrets or mysteries of philosophy; and gave him to understand that himself esteemed it more to excel other men in learning and knowledge than in power and empire. And what use he had of learning doth appear, or rather shine, in all his speeches and answers, being full of science and use of science, and that in all variety.

12.   And herein again it may seem a thing scholastical, and somewhat idle, to recite things that every man knoweth; but yet, since the argument I handle leadeth me thereunto, I am glad that men shall per-

ceive I am as willing to flatter, if they will so call it, an Alexander, or a Caesar, or an Antoninus, that are dead many hundred years since, as any that now liveth: for it is the displaying of the glory of learning in sovereignty that I propound to myself, and not a humour of declaiming in any man's praises. Observe then the speech he used of Diogenes, and see if it tend not to the true state of one of the greatest questions of moral philosophy; whether the enjoying of outward things, or the contemning of them, be the greatest happiness: for when he saw Diogenes so perfectly contented with so little, he said to those that mocked at his condition, *Were I not Alexander, I would wish to be Diogenes.* But Seneca inverteth it, and saith, *Plus erat, quod hic nollet accipere, quam quod ille posset dare.*[139] There were more things which Diogenes would have refused, than those were which Alexander could have given or enjoyed.

13.　Observe again that speech which was usual with him, *That he felt his mortality chiefly in two things, sleep and lust;*[140] and see if it were not a speech extracted out of the depth of natural philosophy, and liker to have come out of the mouth of Aristotle or Democritus than from Alexander.

14.　See again that speech of humanity and poesy; when upon the bleeding of his wounds, he called unto him one of his flatterers, that was wont to ascribe to him divine honour, and said, *Look, this is very blood; this is not such a liquor as Homer speaketh of, which ran from Venus' hand, when it was pierced by Diomedes.*[141]

15.　See likewise his readiness in reprehension of logic, in the speech he used to Cassander, upon a complaint that was made against his father Antipater: for when Alexander happed to say, *Do you think these men would have come from so far to complain, except they had just cause of grief?* and Cassander answered, *Yea, that was the matter, because they thought they should not be disproved;* said Alexander laughing: *See the subtleties of Aristotle, to take a matter both ways, pro et contra, etc.*

16.　But note again how well he could use the same art, which he reprehended, to serve his own humour: when bearing a secret grudge to Callisthenes, because he was against the new ceremony of his adoration, feasting one night where the same Callisthenes was at the table, it was moved by some after supper, for entertainment sake, that Callisthenes, who was an eloquent man, might speak of some theme or purpose at his own choice; which Callisthenes did; choosing the praise of the Macedonian nation for his discourse, and performing the same with so good manner as the hearers were much ravished: whereupon Alexander, nothing pleased, said, *It was easy to be eloquent upon so good a subject:* but, saith he, *Turn your style, and let us hear what you can say against us:* which Callisthenes presently undertook, and did with that sting and life, that Alexander interrupted him and said, *The goodness of the cause made him eloquent before, and despite made him eloquent then again.*

17.   Consider further, for tropes of rhetoric, that excellent use of a metaphor or translation, wherewith he taxed Antipater, who was an imperious and tyrannous governor: for when one of Antipater's friends commended him to Alexander for his moderation, that he did not degenerate, as his other lieutenants did, into the Persian pride, in use of purple, but kept the ancient habit of Macedon, of black;[142] *True,* saith Alexander, *but Antipater is all purple within.*[143] Or that other, when Parmenio came to him in the plain of Arbela, and showed him the innumerable multitude of his enemies, specially as they appeared by the infinite number of lights, as it had bcen a ncw firmament of stars, and thereupon advised him to assail them by night: whereupon he answered, *That he would not steal the victory.*

18.   For matter of policy, weigh that significant distinction, so much in all ages embraced, that he made between his two friends Hephaestion and Craterus, when he said, *That the one loved Alexander, and the other loved the king:* describing the principal difference of princes' best servants, that some in affection love their person, and others in duty love their crown.

19.   Weigh also that excellent taxation of an error, ordinary with counsellors of princes, that they counsel their masters according to the model of their own mind and fortune, and not of their masters'; when upon Darius' great offers Parmenio had said, *Surely I would accept these offers, were I as Alexander;* saith Alexander, *So would I, were I as Parmenio.*

20.   Lastly, weigh that quick and acute reply, which he made when he gave so large gifts to his friends and servants, and was asked what he did reserve for himself, and he answered, *Hope:* weigh, I say, whether

he had not cast up his account aright, because *hope* must be the portion of all that resolve upon great enterprises. For this was Caesar's portion when he went first into Gaul, his estate being then utterly overthrown with largesses. And this was likewise the portion of that noble prince, howsoever transported with ambition, Henry Duke of Guise, of whom it was usually said, that he was the greatest usurer in France, because he had turned all his estate into obligations.

21.   To conclude therefore: as certain critics are used to say hyperbolically, *That if all sciences were lost they might be found in Virgil,* so certainly this may be said truly, there are the prints and footsteps of learning in those few speeches which are reported of this prince: the admiration of whom, when I consider him not as Alexander the Great, but as Aristotle's scholar, hath carried me too far.

22.   As for Julius Caesar, the excellency of his learning needeth not to be argued from his education, or his company, or his speeches; but in a further degree doth declare itself in his writings and works; whereof some are extant and permanent, and some unfortunately perished. For first, we see there is left unto us that excellent history of his own wars, which he entitled only a Commentary, wherein all succeeding times have admired the solid weight of matter, and the real passages and lively images of actions and persons, expressed in the greatest propriety of words and perspicuity of narration that ever was; which that it was not the effect of a natural gift, but of learning and precept, is well witnessed by that work of his entitled *De Analogia,*[144] being a grammatical philosophy, wherein he did labour to make this same *Vox ad placitum* to become *Vox ad licitum,* and to reduce custom of speech to congruity of speech; and took, as it were, the picture of words from the life of reason.

23.   So we receive from him, as a monument both of his power and learning, the then reformed computation of the year; well expressing that he took it to be as great a glory to himself to observe and know the law of the heavens, as to give law to men upon the earth.

24.   So likewise in that book of his, *Anti-Cato,*[145] it may easily appear that he did aspire as well to victory of wit as victory of war: undertaking therein a conflict against the greatest champion with the pen that then lived, Cicero the Orator.

25.   So again in his book of Apophthegms which he collected, we

see that he esteemed it more honour to make himself but a pair of tables to take the wise and pithy words of others, than to have every word of his own to be made an apophthegm or an oracle; as vain princes, by custom of flattery, pretend to do.[146] And yet if I should enumerate divers of his speeches, as I did those of Alexander, they are truly such as Salomon noteth, when he saith, *Verba sapientium tanquam aculei, et tanquam clavi in altum defixi:*[147] whereof I will only recite three, not so delectable for elegancy, but admirable for vigour and efficacy.

26.   As first, it is reason he be thought a master of words, that could with one word appease a mutiny in his army, which was thus: The Romans, when their generals did speak to their army, did use the word *Milites,* but when the magistrates spake to the people, they did use the word *Quirites.* The soldiers were in tumult, and seditiously prayed to be cashiered; not that they so meant, but by expostulation thereof to draw Caesar to other conditions; wherein he being resolute not to give way, after some silence, he began his speech, *Ego Quirites,*[148] which did admit them already cashiered; wherewith they were so surprised, crossed, and confused, as they would not suffer him to go on in his speech, but relinquished their demands, and made it their suit to be again called by the name of *Milites.*

27.   The second speech was thus: Caesar did extremely affect the name of king; and some were set on as he passed by in popular acclamation to salute him king: whereupon, finding the cry weak and poor, he put it off thus, in a kind of jest, as if they had mistaken his surname; *Non Rex sum, sed Caesar;*[149] a speech that if it be searched, the life and fulness of it can scarce be expressed. For first it was a refusal of the name, but yet not serious: again, it did signify an infinite confidence and magnanimity, as if he presumed Caesar was the greater title; as by his worthiness it is come to pass till this day: but chiefly it was a speech of great allurement toward his own purpose; as if the state did strive with him but for a name, whereof mean families were vested; for *Rex* was a surname with the Romans, as well as *King* is with us.[150]

28.   The last speech which I will mention was used to Metellus, when Caesar, after war declared, did possess himself of the city of Rome; at which time entering into the inner treasury to take the money there

accumulate, Metellus being tribune forbade him: whereto Caesar said, *That if he did not desist, he would lay him dead in the place.* And presently taking himself up, he added, *Young man, it is harder for me to speak it than to do it; Adolescens, durius est mihi hoc dicere quam facere.*[151] A speech compounded of the greatest terror and greatest clemency that could proceed out of the mouth of man.[152]

29. But to return and conclude with him, it is evident himself knew well his own perfection in learning, and took it upon him; as appeared when, upon occasion that some spake what a strange resolution it was in Lucius Sulla to resign his dictature; he scoffing at him to his own advantage answered, *That Sulla could not skill of letters, and therefore knew not how to dictate.*[153]

30. And here it were fit to leave this point, touching the concurrence of military virtue and learning (for what example should come with any grace after those two of Alexander and Caesar?), were it not in regard of the rareness of circumstance that I find in one other particular, as that which did so suddenly pass from extreme scorn to extreme wonder; and it is of Xenophon the philosopher, who went from Socrates' school into Asia, in the expedition of Cyrus the younger, against King Artaxerxes. This Xenophon at that time was very young, and never had seen the wars before; neither had any command in the army, but only followed the war as a voluntary, for the love and conversation of Proxenus his friend.[154] He was present when Phalinus came in message from the great king to the Grecians, after that Cyrus was slain in the field, and they a handful of men left to themselves in the midst of the king's territories, cut off from their country by many navigable rivers, and many hundred miles. The message imported that they should deliver up their arms and submit themselves to the king's mercy. To which message before answer was made, divers of the army conferred familiarly with Phalinus, and amongst the rest Xenophon happened to say, *Why, Phalinus, we have now but these two things left, our arms and our virtue; and if we yield up our arms, how shall we make use of our virtue?* Whereto Phalinus smiling on him said, *If I be not deceived, young gentleman, you are an Athenian: and I believe you study philosophy, and it is pretty that you say: but you are much abused, if you think your virtue can withstand the king's power.*[155] Here was the scorn;

the wonder followed: which was, that this young scholar or philosopher, after all the captains were murdered in parley by treason, conducted those ten thousand foot through the heart of all the king's high countries from Babylon to Graecia in safety, in despite of all the king's forces, to the astonishment of the world, and the encouragement of the Grecians in time succeeding to make invasion upon the kings of Persia: as was after purposed by Jason the Thessalian, attempted by Agesilaus the Spartan, and achieved by Alexander the Macedonian, all upon the ground of the act of that young scholar.

VIII

learning

1. To proceed now from imperial and military virtue to moral and private virtue: first, it is an assured truth, which is contained in the verses:

> Scilicet ingenuas didicisse fideliter artes,
> Emollit mores, nec sinit esse feros.[156] *contrary effect*

It taketh away the wildness and barbarism and fierceness of men's minds; but indeed the accent had need be upon *fideliter:* for a little superficial learning doth rather work a contrary effect. It taketh away all levity, temerity, and insolency, by copious suggestion of all doubts and difficulties, and acquainting the mind to balance reasons on both sides, and to turn back the first offers and conceits of the mind, and to accept of nothing but examined and tried. It taketh away vain admiration of anything, which is the root of all weakness: for all things are admired either because they are new, or because they are great. For novelty, no man that wadeth in learning or contemplation thoroughly, but will find that printed in his heart, *Nil novi super terram.*[157] Neither can any man marvel at the play of puppets, that goeth behind the curtain, and adviseth well of the motion. And for magnitude, as Alexander the Great, after that he was used to great armies, and the great conquests of the spacious provinces in Asia, when he received letters out of Greece, of some fights and services there, which were commonly for a passage or a fort, or some walled town at the most, he said, *It seemed to him that he was advertised of the Battle of the Frogs and the Mice, that the old tales went of.*[158] So certainly, if a man meditate much upon the universal frame of nature, the earth with men upon it (the divineness of souls except) will not seem much other than an ant-hill, whereas some ants carry corn, and some carry their young, and some go empty, and all to and fro a little heap of dust. It taketh away or mitigateth fear of death or adverse fortune; which is one of the greatest impediments of virtue, and imperfections of manners. For if a man's mind be deeply seasoned with the consideration of the mortality and corruptible nature of things, he will easily concur with Epictetus, who went forth one day and saw a woman weeping for her pitcher of earth that was broken, and went forth the next day and saw a woman weeping for her son that was dead, and thereupon said: *Heri vidi fragilem frangi, hodie vidi mortalem mori.*[159] And therefore Virgil did excellently and profoundly couple the knowledge of causes and the conquest of all fears, together, as *concomitantia:*

> Felix, qui potuit rerum cognoscere causas,
> Quique metus omnes, et inexorabile fatum
> Subjecit pedibus, strepitumque Acherontis avari.[160]

2.   It were too long to go over the particular remedies which learning doth minister to all the diseases of the mind; sometimes purging the ill humours, sometimes opening the obstructions, sometimes helping digestion, sometimes increasing appetite, sometimes healing the wounds and exulcerations thereof, and the like; and therefore I will conclude with that which hath *rationem totius,* which is, that it disposeth the constitution of the mind not to be fixed or settled in the defects thereof, but still to be capable and susceptible of growth and reformation. For the unlearned man knows not what it is to descend into himself, or to call himself to account; nor the pleasure of that *suavissima vita, indies sentire se fieri meliorem.*[161] The good parts he hath he will learn to show to the full, and use them dexterously, but not much to increase them: the faults he hath he will learn how to hide and colour them, but not much to amend them, like an ill mower, that mows on still, and never whets his scythe. Whereas with the learned man it fares otherwise, that he doth ever intermix the correction and amendment of his mind with the use and employment thereof. Nay further, in general and in sum, certain it is that *Veritas* and *Bonitas* differ but as the seal and the print.

for Truth prints Goodness; and they be the clouds of error which descend in the storms of passions and perturbations.

3.   From moral virtue let us pass on to matter of power and commandment, and consider whether in right reason there be any comparable with that wherewith knowledge investeth and crowneth man's nature. We see the dignity of the commandment is according to the dignity of the commanded: to have commandment over beasts, as herdmen have, is a thing contemptible; to have commandment over children, as schoolmasters have, is a matter of small honour; to have commandment over galley-slaves is a disparagement rather than an honour. Neither is the commandment of tyrants much better, over people which have put off the generosity of their minds: and therefore it was ever holden that honours in free monarchies and commonwealths had a sweetness more than in tyrannies, because the commandment extendeth more over the wills of men, and not only over their deeds and services. And therefore, when Virgil putteth himself forth to attribute to Augustus Caesar the best of human honours, he doth it in these words:

> Victorque volentes
> Per populos dat jura, viamque affectat Olympo.[162]

But yet the commandment of knowledge is yet higher than the commandment over the will: for it is a commandment over the reason, belief, and understanding of man, which is the highest part of the mind, and giveth law to the will itself. For there is no power on earth which setteth up a throne or chair of estate in the spirits and souls of men, and in their cogitations, imaginations, opinions, and beliefs, but knowledge and learning. And therefore we see the detestable and extreme pleasure that arch-heretics, and false prophets, and impostors are transported with, when they once find in themselves that they have a superiority in the faith and conscience of men; so great as if they have once tasted of it, it is seldom seen that any torture or persecution can make them relinquish or abandon it. But as this is that which the author of the Revelation calleth the depth or profoundness of Satan,[163] so by argument of contraries, the just and lawful sovereignty over men's understanding, by force of truth rightly interpreted, is that which approacheth nearest to the similitude of the Divine Rule.

4.   As for fortune and advancement, the beneficence of learning is not so confined to give fortune only to states and commonwealths, as it doth not likewise give fortune to particular persons. For it was well noted long ago, that Homer hath given more men their livings, than either Sulla, or Caesar, or Augustus ever did, notwithstanding their great largesses and donatives, and distributions of lands to so many legions. And no doubt it is hard to say whether arms or learning have advanced greater numbers. And in case of sovereignty we see, that if arms or descent have carried away the kingdom, yet learning hath carried the priesthood, which ever hath been in some competition with empire.[164]

5.   Again, for the pleasure and delight of knowledge and learning, it far surpasseth all other in nature: for, shall the pleasures of the affections so exceed the pleasures of the senses, as much as the obtaining of desire or victory exceedeth a song or a dinner? And must not, of consequence, the pleasures of the intellect or understanding exceed the pleasures of the affections? We see in all other pleasures there is satiety, and after they be used, their verdure departeth; which showeth well they be but deceits of pleasure, and not pleasures: and that it was the novelty which pleased and not the quality; and therefore we see that voluptuous men turn friars, and ambitious princes turn melancholy. But of knowledge there is no satiety, but satisfaction and appetite are perpetually interchangeable; and therefore appeareth to be good in itself simply, without fallacy or accident. Neither is that pleasure of small efficacy and contentment to the mind of man, which the poet Lucretius describeth elegantly,

Suave mari magno, turbantibus aequora ventis, etc.[165]

It is a view of delight, saith he, *to stand or walk upon the shore side, and to see a ship tossed with tempest upon the sea; or to be in a fortified tower, and to see two battles join upon a plain. But it is a pleasure incomparable, for the mind of man to be settled, landed, and fortified in the certainty of truth; and from thence to descry and behold the errors, perturbations, labours, and wanderings up and down of other men.*

6.   Lastly, leaving the vulgar arguments, that by learning man excelleth man in that wherein man excelleth beasts; that by learning man

ascendeth to the heavens and their motions, where in body he cannot come, and the like; let us conclude with the dignity and excellency of knowledge and learning in that whereunto man's nature doth most aspire, which is immortality or continuance: for to this tendeth generation, and raising of houses and families; to this tend buildings, foundations, and monuments; to this tendeth the desire of memory, fame, and celebration; and in effect the strength of all other human desires. We see then how far the monuments of wit and learning are more durable than the monuments of power or of the hands. For have not the verses of Homer continued twenty-five hundred years, or more, without the loss of a syllable or letter; during which time infinite palaces, temples, castles, cities, have been decayed and demolished? It is not possible to have the true pictures or statues of Cyrus, Alexander, Caesar; no, nor of the kings or great personages of much later years; for the originals cannot last, and the copies cannot but leese of the life and truth. But the images of men's wits and knowledges remain in books, exempted from the wrong of time and capable of perpetual renovation. Neither are they fitly to be called images, because they generate still, and cast their seeds in the minds of others, provoking and causing infinite actions and opinions in succeeding ages: so that, if the invention of the ship was thought so noble, which carrieth riches and commodities from place to place, and consociateth the most remote regions in participation of their fruits, how much more are letters to be magnified, which as ships pass through the vast seas of time, and make ages so distant to participate of the wisdom, illuminations, and inventions, the one of the other? Nay further, we see some of the philosophers which were least divine, and most immersed in the senses, and denied generally the immortality of the soul, yet came to this point, that whatsoever motions the spirit of man could act and perform without the organs of the body, they thought might remain after death, which were only those of the understanding, and not of the affection: so immortal and incorruptible a thing did knowledge seem unto them to be. But we, that know by divine revelation that not only the understanding but the affections purified, not only the spirit but the body changed, shall be advanced to immortality, do disclaim in[166] these rudiments of the senses. But it must be remembered, both in this last point and so it may likewise be needful in other places, that in probation of the dignity of knowledge or learning, I did in the beginning separate divine testimony from human, which method I have pursued, and so handled them both apart.

7.   Nevertheless I do not pretend, and I know it will be impossible for me, by any pleading of mine, to reverse the judgment, either of Aesop's Cock, that preferred the barleycorn before the gem; or of Midas, that being chosen judge between Apollo, president of the Muses, and Pan, god of the flocks, judged for plenty;[167] or of Paris, that judged for beauty and love against wisdom and power; or of Agrippina, *occidat matrem, modo imperet,* that preferred empire with any condition never so detestable;[168] or of Ulysses, *qui vetulam praetulit immortalitati,*[169] being a figure of those which prefer custom and habit before all excellency; or of a number of the like popular judgments. For these things must continue as they have been: but so will that also continue whereupon learning hath ever relied, and which faileth not: *Justificata est sapientia a filiis suis.*[170]

THE SECOND BOOK

THE SECOND BOOK OF FRANCIS BACON
OF THE PROFICIENCE AND ADVANCEMENT
OF LEARNING DIVINE AND HUMAN ⌇

To the King

1. It might seem to have more convenience, though it come often otherwise to pass, excellent King, that those which are fruitful in their generations, and have in themselves the foresight of immortality in their descendants, should likewise be more careful of the good estate of future times, unto which they know they must transmit and commend over their dearest pledges. Queen Elizabeth was a sojourner in the world in respect of her unmarried life, and was a blessing to her own times; and yet so as the impression of her good government, besides her happy memory, is not without some effect which doth survive her. But to your Majesty, whom God hath already blessed with so much royal issue, worthy to continue and represent you for ever, and whose youthful and fruitful bed doth yet promise many the like renovations, it is proper and agreeable to be conversant not only in the transitory parts of good government, but in those acts also which are in their nature permanent and perpetual: amongst the which, if affection do not transport me, there is not any more worthy than the further endowment of the world with sound and fruitful knowledge. For why should a few received authors stand up like Hercules' columns,[1] beyond which there should be no sailing or discovering, since we have so bright and benign a star as your Majesty to conduct and prosper us? To return therefore where we left, it remaineth to consider of what kind those acts are which have been undertaken and performed by kings and others for the increase and advancement of learning: wherein I purpose to speak actively without digressing or dilating.

2. Let this ground therefore be laid, that all works are overcome by amplitude of reward, by soundness of direction, and by the conjunction of labours. The first multiplieth endeavour, the second preventeth error, and the third supplieth the frailty of man: but the principal of these is direction: for *Claudus in via antevertit cursorem extra viam;* and Salomon excellently setteth it down, *If the iron be not sharp, it requireth more strength; but wisdom is that which prevaileth;*[2] signifying that the invention or election of the mean is more effectual than any inforcement or accumulation of endeavours. This I am induced to speak, for that (not derogating from the noble intention of any that have been deservers towards the state of learning) I do observe nevertheless that their works and acts are rather matters of magnificence and memory, than of progression and proficience, and tend rather to augment the mass of learning in the multitude of learned men, than to rectify or raise the sciences themselves.

3.   The works or acts of merit towards learning are conversant about three objects: the places of learning, the books of learning, and the persons of the learned. For as water, whether it be the dew of heaven, or the springs of the earth, doth scatter and leese itself in the ground, except it be collected into some receptacle, where it may by union comfort and sustain itself: and for that cause the industry of man hath made and framed spring-heads, conduits, cisterns, and pools, which men have accustomed likewise to beautify and adorn with accomplishments of magnificence and state, as well as of use and necessity: so this excellent liquor of knowledge, whether it descend from divine inspiration, or spring from human sense, would soon perish and vanish to oblivion, if it were not preserved in books, traditions, conferences, and places appointed, as universities, colleges, and schools, for the receipt and comforting of the same.

4.   The works which concern the seats and places of learning are four; foundations and buildings, endowments with revenues, endowments with franchises and privileges, institutions and ordinances for government; all tending to quietness and privateness of life, and discharge of cares and troubles; much like the stations which Virgil prescribeth for the hiving of bees:

> Principio sedes apibus statioque petenda,
> Quo neque sit ventis aditus, etc.[3]

5.   The works touching books are two: first, libraries, which are as the shrines where all the relics of the ancient saints, full of true virtue, and that without delusion or imposture, are preserved and reposed: secondly, new editions of authors, with more correct impressions, more faithful translations, more profitable glosses, more diligent annotations, and the like.

6.   The works pertaining to the persons of learned men, besides the advancement and countenancing of them in general, are two: the reward and designation of readers in sciences already extant and invented; and the reward and designation of writers and inquirers concerning any parts of learning not sufficiently laboured and prosecuted.

7.   These are summarily the works and acts, wherein the merits of many excellent princes and other worthy personages have been conversant. As for any particular commemorations, I call to mind what Cicero said, when he gave general thanks; *Difficile non aliquem, ingratum quenquam praeterire.*[4] Let us rather, according to the Scriptures,[5] look unto that part of the race which is before us, than look back to that which is already attained.

8.   First, therefore, amongst so many great foundations of colleges in Europe, I find strange that they are all dedicated to professions, and none left free to arts and sciences at large. For if men judge that learning should be referred to action, they judge well; but in this they fall into the error described in the ancient fable,[6] in which the other parts of the body did suppose the stomach had been idle, because it neither performed the office of motion, as the limbs do, nor of sense, as the head doth; but yet notwithstanding it is the stomach that digesteth and distributeth to all the rest: so if any man think philosophy and universality to be idle studies, he doth not consider that all professions are from thence served and supplied. And this I take to be a great cause that hath hindered the progression of learning, because these fundamental knowledges have been studied but in passage. For if you will have a tree bear more fruit than it hath used to do, it is not anything you can do to the boughs, but it is the stirring of the earth and putting new mould about the roots that must work it. Neither is it to be forgotten, that this dedicating of foundations and dotations to professory learning hath not only had a malign aspect and influence upon

the growth of sciences, but hath also been prejudicial to states and governments. For hence it proceedeth that princes find a solitude in regard of able men to serve them in causes of estate, because there is no education collegiate which is free; where such as were so disposed might give themselves to histories, modern languages, books of policy and civil discourse, and other the like enablements unto service of estate.

9.   And because Founders of Colleges do plant, and Founders of Lectures do water, it followeth well in order to speak of the defect which is in public lectures; namely, in the smallness and meanness of the salary or reward which in most places is assigned unto them; whether they be lectures of arts, or of professions. For it is necessary to the progression of sciences that Readers be of the most able and sufficient men; as those which are ordained for generating and propagating of sciences, and not for transitory use. This cannot be, except their condition and endowment be such as may content the ablest man to appropriate his whole labour and continue his whole age in that function and attendance; and therefore must have a proportion answerable to that mediocrity or competency of advancement, which may be expected from a profession or the practice of a profession. So as, if you will have sciences flourish, you must observe David's military law, which was, *That those which staid with the carriage should have equal part with those which were in the action;*[7] else will the carriages be ill attended. So Readers in sciences are indeed the guardians of the stores and provisions of sciences, whence men in active courses are furnished, and therefore ought to have equal entertainment with them: otherwise if the fathers in sciences be of the weakest sort or be ill-maintained,

Et patrum invalidi referent jejunia nati.[8]

10.   Another defect I note, wherein I shall need some alchemist to help me, who call upon men to sell their books, and to build furnaces; quitting and forsaking Minerva and the Muses as barren virgins, and relying upon Vulcan.[9] But certain it is, that unto the deep, fruitful, and operative study of many sciences, especially Natural Philosophy and Physic, books be not the only instrumentals; wherein also the beneficence of men hath not been altogether wanting: for we see spheres, globes, astrolabes, maps, and the like, have been provided as

appurtenances to astronomy and cosmography, as well as books: we see likewise that some places instituted for physic have annexed the commodity of gardens for simples of all sorts, and do likewise command the use of dead bodies for anatomies. But these do respect but a few things. In general, there will hardly be any main proficience in the disclosing of nature, except there be some allowance for expenses about experiments; whether they be experiments appertaining to Vulcanus or Daedalus, furnace or engine, or any other kind: and therefore as secretaries and spials of princes and states bring in bills for intelligence, so you must allow the spials and intelligencers of nature to bring in their bills; or else you shall be ill advertised.

11.   And if Alexander made such a liberal assignation to Aristotle of treasure[10] for the allowance of hunters, fowlers, fishers, and the like, that he might compile a History of Nature, much better do they deserve it that travail in Arts of Nature.[11]

12.   Another defect which I note, is an intermission or neglect, in those which are governors in universities, of consultation; and in princes or superior persons, of visitation: to enter into account and consideration, whether the readings, exercises, and other customs appertaining unto learning, anciently begun and since continued, be well instituted or no; and thereupon to ground an amendment or reformation in that which shall be found inconvenient. For it is one of your Majesty's own most wise and princely maxims, *That in all usages and precedents, the times be considered wherein they first began; which, if they were weak or ignorant, it derogateth from the authority of the usage, and leaveth it for suspect.* And therefore inasmuch as most of the usages and orders of the universities were derived from more obscure times, it is the more requisite they be re-examined. In this kind I will give an instance or two, for example sake, of things that are the most obvious and familiar. The one is a matter, which though it be ancient and general, yet I hold to be an error; which is, that scholars in universities come too soon and too unripe to logic and rhetoric, arts fitter for graduates than children and novices: for these two, rightly taken, are the gravest of sciences, being the arts of arts; the one for judgment, the other for ornament: and they be the rules and directions how to set forth and dispose matter; and therefore for minds empty and unfraught with matter, and which have not gathered that which Cicero calleth *Sylva* and *Supellex*,[12] stuff and variety, to begin with those arts (as if one should learn to weigh, or to measure, or to paint the wind) doth work but this effect, that the wisdom of those arts, which is great and universal, is almost made contemptible, and is degenerate into childish sophistry and ridiculous affectation. And further, the untimely learning of them hath drawn on by consequence the superficial and unprofitable teaching and writing of them, as fitteth indeed to the capacity of children. Another is a lack I find in the exercises used in the Universities, which do make too great a divorce between invention and memory; for their speeches are either premeditate in *verbis conceptis,* where nothing is left to invention, or merely extemporal, where little is left to memory: whereas in life and action there is least use of either of these, but rather of intermixtures of premeditation and invention, notes and memory; so as the exercise fitteth not the practice, nor the image the life; and it is ever a true rule in exercises, that they be framed as near as may be to the life of practice; for otherwise they do pervert the motions and faculties of the mind, and not prepare them. The truth whereof is not obscure, when scholars come to the practices of professions, or other actions of civil life; which when they set into, this want is soon found by themselves, and sooner by others. But this part, touching the amendment of the institutions and orders of Universities, I will conclude with the clause of Caesar's letter to Oppius and Balbus, *Hoc quemadmodum fieri possit, nonnulla mihi in mentem veniunt, et multa reperiri possunt; de iis rebus rogo vos ut cogitationem suscipiatis.*[13]

13.   Another defect which I note, ascendeth a little higher than the precedent: for as the proficience of learning consisteth much in the orders and institutions of Universities in the same states and kingdoms, so it would be yet more advanced, if there were more intelligence mutual between the Universities of Europe than now there is. We see there may be many orders and foundations, which though they be divided under several sovereignties and territories, yet they take themselves to have a kind of contract, fraternity, and correspondence one with the other, insomuch as they have provincials and generals. And surely as nature createth brotherhood in families, and arts mechanical contract brotherhoods in communalities, and the anointment of God super-

induceth a brotherhood in kings and bishops, so in like manner there cannot but be a fraternity in learning and illumination, relating to that paternity which is attributed to God, who is called the Father of illuminations or lights.[14]

14. The last defect which I will note is, that there hath not been, or very rarely been, any public designation of writers or inquirers concerning such parts of knowledge as may appear not to have been already sufficiently laboured or undertaken; unto which point it is an inducement to enter into a view and examination what parts of learning have been prosecuted and what omitted: for the opinion of plenty is amongst the causes of want, and the great quantity of books maketh a show rather of superfluity than lack; which surcharge nevertheless is not to be remedied by making no more books, but by making more good books, which, as the serpent of Moses, might devour the serpents of the enchanters.[15]

15. The removing of all the defects formerly enumerated, except the last, and of the active part also of the last (which is the designation of writers), are *opera basilica;* towards which the endeavours of a private man may be but as an image in a crossway, that may point at the way, but cannot go it: but the inducing part of the latter, which is the survey of learning, may be set forward by private travail. Wherefore I will now attempt to make a general and faithful perambulation of learning, with an inquiry what parts thereof lie fresh and waste, and not improved and converted by the industry of man; to the end that such a plot made and recorded to memory, may both minister light to any public designation, and also serve to excite voluntary endeavours: wherein nevertheless my purpose is at this time to note only omissions and deficiencies, and not to make any redargution of errors or incomplete prosecutions; for it is one thing to set forth what ground lieth unmanured, and another thing to correct ill husbandry in that which is manured.

In the handling and undertaking of which work I am not ignorant what it is that I do now move and attempt, nor insensible of mine own weakness to sustain my purpose; but my hope is, that if my extreme love to learning carry me too far, I may obtain the excuse of affection; for that *It is not granted to man to love and to be wise.*[16] But I know well I can use no other liberty of judgment than I must leave to others; and

I for my part shall be indifferently glad either to perform myself, or accept from another, that duty of humanity; *Nam qui erranti comiter monstrat viam, etc.*[17] I do foresee likewise that of those things which I shall enter and register as deficiencies and omissions, many will conceive and censure that some of them are already done and extant; others to be but curiosities, and things of no great use; and others to be of too great difficulty, and almost impossibility to be compassed and effected. But for the two first, I refer myself to the particulars; for the last, touching impossibility, I take it those things are to be held possible which may be done by some person, though not by every one; and which may be done by many, though not by any one; and which may be done in succession of ages, though not within the hour-glass of one man's life; and which may be done by public designation, though not by private endeavour. But notwithstanding, if any man will take to himself rather that of Salomon, *Dicit piger, Leo est in via,*[18] than that of Virgil, *Possunt quia posse videntur,*[19] I shall be content that my labours be esteemed but as the better sort of wishes: for as it asketh some knowledge to demand a question not impertinent, so it requireth some sense to make a wish not absurd.

I 1. The *parts* of human learning have reference to the three parts of man's understanding, which is the seat of learning: *history* to his *memory,* *poesy to his imagination,* and *philosophy* to his *reason.* Divine learning receiveth the same distribution; for the spirit of man is the same, though the revelation of oracle and sense be diverse: so as theology consisteth also of *history* of the church; of *parables, which is divine poesy;* and of holy *doctrine* or *precept:* for as for that part which seemeth supernumerary, which is *prophecy,* it is but Divine History; which hath that prerogative over human, as the narration may be before the fact as well as after.

2. History is natural, civil, ecclesiastical, and literary; whereof the first three I allow as extant, the fourth I note as deficient. For no man hath propounded to himself the general state of learning to be described and represented from age to age, as many have done the works of nature, and the state civil and ecclesiastical; without which the his-

tory of the world seemeth to me to be as the statua of Polyphemus with his eye out; that part being wanting which doth most show the spirit and life of the person: and yet I am not ignorant that in divers particular sciences, as of the jurisconsults, the mathematicians, the rhetoricians, the philosophers, there are set down some small memorials of the schools, authors, and books; and so likewise some barren relations touching the invention of arts or usages. But a just story of learning, containing the antiquities and originals of knowledges and their sects, their inventions, their traditions, their diverse administrations and managings, their flourishings, their oppositions, decays, depressions, oblivions, removes, with the causes and occasions of them, and all other events concerning learning, throughout the ages of the world, I may truly affirm to be wanting. The use and end of which work I do not so much design for curiosity or satisfaction of those that are the lovers of learning, but chiefly for a more serious and grave purpose; which is this in few words, that it will make learned men wise in the use and administration of learning. For it is not St. Augustine's nor St. Ambrose's works that will make so wise a divine, as ecclesiastical history, thoroughly read and observed; and the same reason is of learning.

3.   *History of nature* is of three sorts: of *nature* in *course*, of *nature erring* or *varying*, and of *nature altered* or wrought; that is, *history of creatures, history of marvels,* and *history of arts.* The first of these, no doubt, is extant, and that in good perfection; the two latter are handled so weakly and unprofitably, as I am moved to note them as deficient. For I find no sufficient or competent collection of the works of nature which have a digression and deflection from the ordinary course of generations, productions, and motions; whether they be singularities of place and region, or the strange events of time and chance, or the effects of yet unknown properties, or the instances of exception to general kinds. It is true, I find a number of books of fabulous experiments and secrets, and frivolous impostures for pleasure and strangeness; but a substantial and severe collection of the *heteroclites* or *irregulars* of *nature,*[20] well examined and described, I find not: especially not with due rejection of fables and popular errors: for as things now are, if an untruth in nature be once on foot, what by reason of the neglect of examination and countenance of antiquity, and what by reason of the use of the

opinion in similitudes and ornaments of speech, it is never called down.

4.   The use of this work, honoured with a precedent in Aristotle,[21] is nothing less than to give contentment to the appetite of curious and vain wits, as the manner of Mirabilaries[22] is to do; but for two reasons, both of great weight; the one to correct the partiality of axioms and opinions, which are commonly framed only upon common and familiar examples; the other because from the wonders of nature is the nearest intelligence and passage towards the wonders of art: for it is no more but by following, and as it were hounding nature in her wanderings, to be able to lead her afterwards to the same place again. Neither am I of opinion, in this history of marvels, that superstitious narrations of sorceries, witchcrafts, dreams, divinations, and the like, where there is an assurance and clear evidence of the fact, be altogether excluded. For it is not yet known in what cases and how far effects attributed to superstition do participate of natural causes: and therefore howsoever the practice of such things is to be condemned, yet from the speculation and consideration of them light may be taken, not only for the discerning of the offences, but for the further disclosing of nature. Neither ought a man to make scruple of entering into these things for inquisition of truth, as your Majesty hath showed in your own example; who with the two clear eyes of religion and natural philosophy have looked deeply and wisely into these shadows, and yet proved yourself to be of the nature of the sun, which passeth through pollutions and itself remains as pure as before.[23] But this I hold fit, that these narrations, which have mixture with superstition, be sorted by themselves, and not be mingled with the narrations which are merely and sincerely natural. But as for the narrations touching the prodigies and miracles of religions, they are either not true, or not natural; and therefore impertinent for the story of nature.

5.   For *history of nature wrought* or *mechanical*, I find some collections made of agriculture, and likewise of manual arts; but commonly with a rejection of experiments familiar and vulgar. For it is esteemed a kind of dishonour unto learning to descend to inquiry or meditation upon matters mechanical, except they be such as may be thought secrets, rarities, and special subtleties; which humour of vain and supercilious arrogancy is justly derided in Plato; where he brings in Hip-

pias, a vaunting sophist, disputing with Socrates, a true and unfeigned inquisitor of truth; where the subject being touching beauty, Socrates, after his wandering manner of inductions, put first an example of a fair virgin, and then of a fair horse, and then of a fair pot well glazed, whereat Hippias was offended, and said, *More than for courtesy's sake, he did think much to dispute with any that did allege such base and sordid instances:* whereunto Socrates answereth, *You have reason, and it becomes you well, being a man so trim in your vestiments, etc.,* and so goeth on in an irony.[24] But the truth is, they be not the highest instances that give the securest information; as may be well expressed in the tale so common of the philosopher,[25] that while he gazed upwards to the stars fell into the water; for if he had looked down he might have seen the stars in the water, but looking aloft he could not see the water in the stars. So it cometh often to pass, that mean and small things discover great, better than great can discover the small: and therefore Aristotle noteth well, *That the nature of everything is best seen in his smallest portions.* And for that cause he inquireth the nature of a commonwealth, first in a family, and the simple conjugations of man and wife, parent and child, master and servant, which are in every cottage.[26] Even so likewise the nature of this great city of the world, and the policy thereof, must be first sought in mean concordances and small portions. So we see how that secret of nature, of the turning of iron touched with the loadstone towards the north, was found out in needles of iron, not in bars of iron.

6. But if my judgment be of any weight, the use of *history mechanical* is of all others the most radical and fundamental towards natural philosophy; such natural philosophy as shall not vanish in the fume of subtle, sublime, or delectable speculation, but such as shall be operative to the endowment and benefit of man's life: for it will not only minister and suggest for the present many ingenious practices in all trades, by a connection and transferring of the observations of one art to the use of another, when the experiences of several mysteries shall fall under the consideration of one man's mind; but further, it will give a more true and real illumination concerning causes and axioms than is hitherto attained. For like as a man's disposition is never well known till he be crossed, nor Proteus ever changed shapes till he was straitened

and held fast;[27] so the passages and variations of nature cannot appear so fully in the liberty of nature as in the trials and vexations of art.

II   **1.**   For *civil history,* it is of three kinds; not unfitly to be compared with the three kinds of pictures or images: for of picture or images, we see some are unfinished, some are perfect, and some are defaced. So of histories we may find three kinds, *memorials, perfect histories,* and *antiquities;* for *memorials* are history unfinished, or the first or rough draughts of history; and *antiquities* are history defaced, or some remnants of history which have casually escaped the shipwreck of time.

2.   *Memorials,* or *preparatory history,* are of two sorts; whereof the one may be termed *commentaries,* and the other *registers. Commentaries* are they which set down a continuance of the naked events and action, without the motives or designs, the counsels, the speeches, the pretexts, the occasions and other passages of action: for this is the true nature of a commentary; though Caesar, in modesty mixed with greatness, did for his pleasure apply the name of a commentary to the best history of the world. *Registers* are collections of public acts, as decrees of council, judicial proceedings, declarations and letters of estate, orations and the like, without a perfect continuance or contexture of the thread of the narration.

3.   *Antiquities,* or remnants of history, are, as was said, *Tanquam tabula naufragii:*[28] when industrious persons, by an exact and scrupulous diligence and observation, out of monuments, names, words, proverbs, traditions, private records and evidences, fragments of stories, passages of books that concern not story,[29] and the like, do save and recover somewhat from the deluge of time.

4.   In these kinds of unperfect histories I do assign no deficience, for they are *Tanquam imperfecte mista;* and therefore any deficience in them is but their nature. As for the corruptions and moths of history, which are *epitomes,* the use of them deserveth to be banished, as all men of sound judgment have confessed, as those that have fretted and corroded the sound bodies of many excellent histories, and wrought them into base and unprofitable dregs.[30]

5.   History, which may be called *just* and *perfect* history, is of three kinds, according to the object which it propoundeth or pretendeth to represent: for it either representeth a *time,* or a *person,* or an *action.* The first we call *chronicles,* the second *lives,* and the third *narrations* or *relations.* Of these, although the first be the most complete and absolute kind of history, and hath most estimation and glory, yet the second excelleth it in profit and use, and the third in verity and sincerity. For history of *times* representeth the magnitude of actions, and the public faces and deportments of persons, and passeth over in silence the smaller passages and motions of men and matters. But such being the workmanship of God, as He doth hang the greatest weight upon the smallest wires, *Maxima e minimis suspendens,*[31] it comes therefore to pass, that such histories do rather set forth the pomp of business than the true and inward resorts thereof. But *lives,* if they be well written, propounding to themselves a person to represent in whom actions both greater and smaller, public and private, have a commixture, must of necessity contain a more true, native, and lively representation. So again *narrations* and relations of actions, as the war of Peloponnesus, the expedition of Cyrus Minor, the conspiracy of Catiline, cannot but be more purely and exactly true than histories of times, because they may choose an argument comprehensible within the notice and instructions of the writer: whereas he that undertaketh the story of a time, especially of any length, cannot but meet with many blanks and spaces which he must be forced to fill up out of his own wit and conjecture.

6.   For the *history of times,* I mean of *civil history,* the providence of God hath made the distribution: for it hath pleased God to ordain and illustrate two exemplar states of the world for arms, learning, moral virtue, policy, and laws; the state of Graecia and the state of Rome; the histories whereof, occupying the *middle part* of time, have more ancient to them histories which may by one common name be termed the *antiquities* of the world: and after them histories which may be likewise called by the name of *modern history.*

7.   Now to speak of the deficiencies. As to the *heathen antiquities* of the world, it is in vain to note them for deficient: deficient they are no doubt, consisting most of fables and fragments; but the deficience cannot be holpen; for antiquity is like fame, *Caput inter nubila condit,*[32] her head is muffled from our sight. For the history of the *exemplar*

states, it is extant in good perfection. Not but I could wish there were a perfect course of history for Graecia from Theseus to Philopoemen (what time the affairs of Graecia drowned and extinguished in the affairs of Rome); and for Rome from Romulus to Justinianus, who may be truly said to be *Ultimus Romanorum.*[33] In which sequences of story the text of Thucydides and Xenophon in the one, and the texts of Livius, Polybius, Sallustius, Caesar, Appianus, Tacitus, Herodianus in the other, to be kept entire without any diminution at all, and only to be supplied and continued. But this is a matter of magnificence, rather to be commended than required: and we speak now of parts of learning supplemental and not of supererogation.

8.   But for *modern histories,* whereof there are some few very worthy, but the greater part beneath mediocrity (leaving the care of foreign stories to foreign states, because I will not be *curiosus in aliena republica*[34]), I cannot fail to represent to your Majesty the unworthiness of the history of England in the main continuance thereof, and the partiality and obliquity of that of Scotland in the latest and largest author that I have seen:[35] supposing that it would be honour for your Majesty, and a work very memorable, if this island of Great Britain, as it is now joined in monarchy for the ages to come, so were joined in one history for the times passed; after the manner of the Sacred History, which draweth down the story of the ten tribes and of the two tribes as twins together. And if it shall seem that the greatness of this work may make it less exactly performed, there is an excellent period of a much smaller compass of time, as to the story of England; that is to say, from the uniting of the Roses to the uniting of the kingdoms; a portion of time wherein, to my understanding, there hath been the rarest varieties that in like number of successions of any hereditary monarchy hath been known. For it beginneth with the mixed adoption of a crown by arms and title: an entry by battle, an establishment by marriage, and therefore times answerable, like waters after a tempest, full of working and swelling, though without extremity of storm; but well passed through by the wisdom of the pilot, being one of the most sufficient kings of all the number. Then followeth the reign of a king, whose actions, howsoever conducted, had much intermixture with the affairs of Europe, balancing and inclining them variably; in whose time also began that

great alteration in the state ecclesiastical, an action which seldom cometh upon the stage. Then the reign of a minor: then an offer of a usurpation, though it was but as *febris ephemera*. Then the reign of a queen matched with a foreigner: then of a queen that lived solitary and unmarried, and yet her government so masculine, as it had greater impression and operation upon the states abroad than it any ways received from thence. And now last, this most happy and glorious event, that this island of Britain, divided from all the world,[36] should be united in itself: and that oracle of rest, given to Aeneas, *antiquam exquirite matrem,*[37] should now be performed and fulfilled upon the nations of England and Scotland, being now reunited in the ancient mother name of Britain, as a full period of all instability and peregrinations. So that as it cometh to pass in massive bodies, that they have certain trepidations and waverings before they fix and settle, so it seemeth that by the providence of God this monarchy, before it was to settle in your Majesty and your generations (in which I hope it is now established for ever), had these prelusive changes and varieties.

9. For *lives,* I do find it strange that these times have so little esteemed the virtues of the times, as that the writings of lives should be no more frequent. For although there be not many sovereign princes or absolute commanders, and that states are most collected into monarchies, yet are there many worthy personages that deserve better than dispersed report or barren elogies. For herein the invention of one of the late poets[38] is proper, and doth well enrich the ancient fiction: for he feigneth that at the end of the thread or web of every man's life there was a little medal containing the person's name, and that Time waited upon the shears; and as soon as the thread was cut, caught the medals, and carried them to the river of Lethe; and about the bank there were many birds flying up and down, that would get the medals and carry them in their beak a little while, and then let them fall into the river: only there were a few swans, which if they got a name, would carry it to a temple where it was consecrate. And although many men, more mortal in their affections than in their bodies, do esteem desire of name and memory but as a vanity and ventosity,

Animi nil magnae laudis egentes;[39]

which opinion cometh from that root, *Non prius laudes contempsimus, quam laudanda facere desivimus:*[40] yet that will not alter Salomon's judgment, *Memoria justi cum laudibus, at impiorum nomen putrescet:*[41] the one flourisheth, the other either consumeth to present oblivion, or turneth to an ill odour. And therefore in that style or addition, which is and hath been long well received and brought in use, *Felicis memoriae, piae memoriae, bonae memoriae,* we do acknowledge that which Cicero saith, borrowing it from Demosthenes, that *Bona fama propria possessio defunctorum;*[42] which possession I cannot but note that in our times it lieth much waste, and that therein there is a deficience.

10.   For *narrations* and *relations* of particular actions, there were also to be wished a greater diligence therein; for there is no great action but hath some good pen which attends it. And because it is an ability not common to write a good history, as may well appear by the small number of them; yet if particularity of actions memorable were but tolerably reported as they pass, the compiling of a complete history of times might be the better expected, when a writer should arise that were fit for it: for the collection of such relations might be as a nursery garden, whereby to plant a fair and stately garden, when time should serve.

11.   There is yet another portion of history which Cornelius Tacitus maketh, which is not to be forgotten, specially with that application which he accoupleth it withal, *annals* and *journals:* appropriating to the former matters of estate, and to the latter acts and accidents of a meaner nature. For giving but a touch of certain magnificent buildings, he addeth, *Cum ex dignitate populi Romani repertum sit, res illustres annalibus talia diurnis urbis actis mandare.*[43] So as there is a kind of contemplative heraldry, as well as civil. And as nothing doth derogate from the dignity of a state more than confusion of degrees, so it doth not a little imbase the authority of a history, to intermingle matters of triumph, or matters of ceremony, or matters of novelty, with matters of state. But the use of a *journal* hath not only been in the history of times, but likewise in the history of persons, and chiefly of actions; for princes in ancient time had, upon point of honour and policy both, journals kept, what passed day by day: for we see the chronicle which was read before Ahasuerus,[44] when he could not take rest, contained matter of affairs indeed, but such as had passed in his own time, and

very lately before: but the journal of Alexander's house expressed every small particularity, even concerning his person and court;[45] and it is yet a use well received in enterprises memorable, as expeditions of war, navigations, and the like, to keep *diaries* of that which passeth continually.

12. I cannot likewise be ignorant of a form of writing which some grave and wise men have used, containing a scattered history of those actions which they have thought worthy of memory, with politic discourse and observation thereupon: not incorporate into the history, but separately, and as the more principal in their intention;[46] which kind of *ruminated history* I think more fit to place amongst books of policy, whereof we shall hereafter speak, than amongst books of history: for it is the true office of history to represent the events themselves together with the counsels, and to leave the observations and conclusions thereupon to the liberty and faculty of every man's judgment. But mixtures are things irregular, whereof no man can define.

13. So also is there another kind of history manifoldly mixed, and that is *history of cosmography:* being compounded of natural history, in respect of the regions themselves; of history civil, in respect of the habitations, regiments, and manners of the people; and the *mathematics,* in respect of the climates and configurations towards the heavens: which part of learning of all others in this latter time hath obtained most proficience. For it may be truly affirmed to the honour of these times, and in a virtuous emulation with antiquity, that this great building of the world had never through-lights made in it, till the age of us and our fathers: for although they had knowledge of the Antipodes,

Nosque ubi primus equis Oriens afflavit anhelis,
Illic sera rubens accendit lumina Vesper:[47]

yet that might be by demonstration, and not in fact; and if by travel, it requireth the voyage but of half the globe. But to circle the earth, as the heavenly bodies do, was not done or enterprised till these later times: and therefore these times may justly bear in their word, not only *plus ultra,*[48] in precedence of the ancient *non ultra,* and *imitabile fulmen,* in precedence of the ancient *non imitabile fulmen,*

Demens qui nimbos et non imitabile fulmen, etc.[49]

but likewise *imitabile caelum;* in respect of the many memorable voyages after the manner of heaven about the globe of the earth.

14. And this proficience in navigation and discoveries may plant also an expectation of the further proficience and augmentation of all sciences; because it may seem they are ordained by God to be coevals, that is, to meet in one age. For so the prophet Daniel, speaking of the latter times, foretelleth, *Plurimi pertransibunt, et multiplex erit scientia:*[50] as if the openness and through-passage of the world and the increase of knowledge were appointed to be in the same ages; as we see it is already performed in great part; the learning of these later times not much giving place to the former two periods or returns of learning, the one of the Grecians, the other of the Romans.

III 1. *History ecclesiastical* receiveth the same divisions with history civil: but further, in the propriety thereof, may be divided into the *history of the church,* by a general name; *history of prophecy;* and *history of providence.* The first describeth the times of the militant church, whether it be fluctuant, as the ark of Noah, or moveable, as the ark in the wilderness, or at rest, as the ark in the temple: that is, the state of the church in persecution, in remove, and in peace. This part I ought in no sort to note as deficient; only I would the virtue and sincerity of it were according to the mass and quantity. But I am not now in hand with censures, but with omissions.

2. The second, which is *history of prophecy,* consisteth of two relatives, the prophecy, and the accomplishment; and therefore the nature of such a work ought to be, that every prophecy of the Scripture be sorted with the event fulfilling the same, throughout the ages of the world; both for better confirmation of faith, and for the better illumination of the Church touching those parts of prophecies which are yet unfulfilled: allowing nevertheless that latitude which is agreeable and familiar unto divine prophecies; being of the nature of their Author, with whom a thousand years are but as one day;[51] and therefore are not fulfilled punctually at once, but have springing and germinant accomplishment throughout many ages; though the height or fulness of them may refer to some one age. This is a work which I find deficient; but is to be done with wisdom, sobriety, and reverence, or not at all.

3.   The third, which is *history of providence,* containeth that excellent correspondence which is between God's revealed will and His secret will: which though it be so obscure, as for the most part it is not legible to the natural man; no, nor many times to those that behold it from the Tabernacle; yet at some times it pleaseth God, for our better establishment and the confuting of those which are as without God in the world, to write it in such text and capital letters, that as the prophet saith, *He that runneth by may read it;*[52] that is, mere sensual persons, which hasten by God's judgments, and never bend or fix their cogitations upon them, are nevertheless in their passage and race urged to discern it. Such are the notable events and examples of God's judgments, chastisements, deliverances, and blessings: and this is a work which hath passed through the labour of many, and therefore I cannot present as omitted.

4.   There are also other parts of learning which are *appendices to history:* for all the exterior proceedings of man consist of words and deeds: whereof history doth properly receive and retain in memory the deeds, and if words, yet but as inducements and passages to deeds: so are there other books and writings, which are appropriate to the custody and receipt of words only; which likewise are of three sorts: *orations, letters,* and *brief speeches* or *sayings.* Orations are pleadings, speeches of counsel, laudatives, invectives, apologies, reprehensions, orations of formality or ceremony, and the like. Letters are according to all the variety of occasions, advertisements, advices, directions, propositions, petitions, commendatory, expostulatory, satisfactory, of compliment, of pleasure, of discourse, and all other passages of action. And such as are written from wise men, are of all the words of man, in my judgment, the best; for they are more natural than orations and public speeches, and more advised than conferences or present speeches. So again letters of affairs from such as manage them, or are privy to them, are of all others the best instructions for history, and to a diligent reader the best histories in themselves. For Apophthegms, it is a great loss of that book of Caesar's;[53] for as his history, and those few letters of his which we have, and those apophthegms which were of his own, excel all men's else, so

I suppose would his collection of Apophthegms have done; for as for those which are collected by others, either I have no taste in such matters, or else their choice hath not been happy. But upon these three kinds of writings I do not insist, because I have no deficiencies to propound concerning them.

5.   Thus much therefore concerning history; which is that part of learning which answereth to one of the cells, domiciles, or offices of the mind of man: which is that of the memory.

IV

1.   *Poesy* is a part of learning in measure of words for the most part restrained, but in all other points extremely licensed, and doth truly refer to the imagination; which, being not tied to the laws of matter, may at pleasure join that which nature hath severed, and sever that which nature hath joined; and so make unlawful matches and divorces of things; *Pictoribus atque poetis, etc.*[54] It is taken in two senses in respect of words or matter; in the first sense it is but a *character* of style, and belongeth to arts of speech, and is not pertinent for the present: in the latter it is, as hath been said, one of the principal portions of learning, and is nothing else but *feigned history,* which may be styled as well in prose as in verse.

2.   The use of this *feigned history* hath been to give some shadow of satisfaction to the mind of man in those points wherein the nature of things doth deny it, the world being in proportion inferior to the soul; by reason whereof there is, agreeable to the spirit of man, a more ample greatness, a more exact goodness, and a more absolute variety, than can be found in the nature of things. Therefore, because the acts or events of *true history* have not that magnitude which satisfieth the mind of man, *poesy* feigneth acts and events greater and more heroical: because *true history* propoundeth the successes and issues of actions not so agreeable to the merits of virtue and vice, therefore poesy feigns them more just in retribution, and more according to revealed providence: because true history representeth actions and events more ordinary and less interchanged, therefore poesy endueth them with more rareness, and more unexpected and alternative variations: so as it appeareth that poesy serveth and conferreth to magnanimity, morality,

and to delectation. And therefore it was ever thought to have some participation of divineness, because it doth raise and erect the mind, by submitting the shows of things to the desires of the mind; whereas reason doth buckle and bow the mind into the nature of things. And we see that by these insinuations and congruities with man's nature and pleasure, joined also with the agreement and comfort it hath with music, it hath had access and estimation in rude times and barbarous regions, where other learning stood excluded.

3.   The division of poesy which is aptest in the propriety thereof (besides those divisions which are common unto it with history, as feigned chronicles, feigned lives, and the appendices of history, as feigned epistles, feigned orations, and the rest) is into *poesy narrative, representative,* and *allusive.* The *Narrative* is a mere imitation of history, with the excesses before remembered; choosing for subject commonly wars and love, rarely state, and sometimes pleasure or mirth. *Representative* is as a visible history; and is an image of actions as if they were present, as history is of actions in nature as they are, (that is) past. *Allusive or Parabolical* is a *narration* applied only to express some special purpose or conceit. Which latter kind of parabolical wisdom was much more in use in the ancient times, as by the fables of Aesop, and the brief sentences of the Seven, and the use of hieroglyphics may appear. And the cause was, for that it was then of necessity to express any point of reason which was more sharp or subtle than the vulgar in that manner, because men in those times wanted both variety of examples and subtlety of conceit: and as hieroglyphics were before letters, so parables were before arguments: and nevertheless now and at all times they do retain much life and vigour; because reason cannot be so sensible, nor examples so fit.

4.   But there remaineth yet another use of poesy parabolical, opposite to that which we last mentioned: for that tendeth to demonstrate and illustrate that which is taught or delivered, and this other to retire and obscure it: that is, when the secrets and mysteries of religion, policy, or philosophy, are involved in fables or parables. Of this in divine poesy we see the use is authorized. In heathen poesy we see the exposition of fables doth fall out sometimes with great felicity; as in the fable that the giants being overthrown in their war against the gods, the Earth their mother in revenge thereof brought forth Fame:

Illam terra parens, ira irritata Deorum,
Extremam, ut perhibent, Coeo Enceladoque sororem
Progenuit:[55]   *Aen 4,178*

expounded that when princes and monarchs have suppressed actual and open rebels, then the malignity of the people, which is the mother of rebellion, doth bring forth libels and slanders, and taxations of the states, which is of the same kind with rebellion, but more feminine. So in the fable, that the rest of the gods having conspired to bind Jupiter, Pallas[56] called Briareus with his hundred hands to his aid: expounded that monarchies need not fear any curbing of their absoluteness by mighty subjects, as long as by wisdom they keep the hearts of the people, who will be sure to come in on their side. So in the fable that Achilles was brought up under Chiron the Centaur, who was part a man and part a beast, expounded ingeniously but corruptly by Machiavel,[57] that it belongeth to the education and discipline of princes to know as well how to play the part of the lion in violence, and the fox in guile, as of the man in virtue and justice. Nevertheless, in many the like encounters, I do rather think that the fable was first, and the exposition devised, than that the moral was first, and thereupon the fable framed. For I find it was an ancient vanity in Chrysippus, that troubled himself with great contention to fasten the assertions of the Stoics upon the fictions of the ancient poets; but yet that all the fables and fictions of the poets were but pleasure and not figure, I interpose no opinion. Surely of those poets which are now extant, even Homer himself (notwithstanding he was made a kind of Scripture by the latter schools of the Grecians), yet I should without any difficulty pronounce that his fables had no such inwardness in his own meaning; but what they might have upon a more original tradition, is not easy to affirm; for he was not the inventor of many of them.[58]

5.   In this third[59] part of learning, which is poesy, I can report no deficience. For being as a plant that cometh of the lust of the earth, without a formal seed, it hath sprung up and spread abroad more than any other kind. But to ascribe unto it that which is due, for the expressing of affections, passions, corruptions, and customs, we are beholding to poets more than to the philosophers' works; and for wit and eloquence, not much less than to orators' harangues. But it is not

good to stay too long in the theatre. Let us now pass on to the judicial place or palace of the mind, which we are to approach and view with more reverence and attention.

V 1. The knowledge of man is as the waters, some descending from above, and some springing from beneath; the one informed by the light of nature, the other inspired by divine revelation. The light of nature consisteth in the notions of the mind and the reports of the senses: for as for knowledge which man receiveth by teaching, it is cumulative and not original; as in a water that besides his own spring-head is fed with other springs and streams. So then, according to these two differing illuminations or originals, knowledge is first of all divided into *divinity* and *philosophy*.

2. In *Philosophy*, the contemplations of man do either penetrate unto God, or are circumferred to nature, or are reflected or reverted upon himself. Out of which several inquiries there do arise three knowledges: *divine philosophy, natural philosophy,* and *human philosophy* or *humanity*. For all things are marked and stamped with this triple character of the power of God, the difference of nature, and the use of man. But because the distributions and partitions of knowledge are not like several lines that meet in one angle, and so touch but in a point; but are like branches of a tree, that meet in a stem, which hath a dimension and quantity of entireness and continuance, before it come to discontinue and break itself into arms and boughs: therefore it is good, before we enter into the former distribution, to erect and constitute one universal science, by the name of *philosophia prima, primitive* or *summary philosophy*, as the main and common way, before we come where the ways part and divide themselves; which science whether I should report as deficient or no, I stand doubtful. For I find a certain rhapsody of natural theology, and of divers parts of logic; and of that part of natural philosophy which concerneth the principles, and of that other part of natural philosophy which concerneth the soul or spirit; all these strangely commixed and confused; but being examined, it seemeth to me rather a depredation of other sciences, advanced and exalted unto some height of terms, than anything solid or sub-

stantive of itself. Nevertheless I cannot be ignorant of the distinction which is current, that the same things are handled but in several respects. As for example, that logic considereth of many things as they are in notion, and this philosophy as they are in nature; the one in appearance, the other in existence; but I find this difference better made than pursued. For if they had considered *quantity, similitude, diversity,* and the rest of those extern characters of things, as philosophers, and in nature, their inquiries must of force have been of a far other kind than they are. For doth any of them, in handling *quantity,* speak of the force of union, how and how far it multiplieth virtue? Doth any give the reason, why some things in nature are so common, and in so great mass, and others so rare, and in so small quantity? Doth any, in handling *similitude* and *diversity,* assign the cause why iron should not move to iron, which is more like, but move to the lode-stone, which is less like? Why in all diversities of things there should be certain *participles* in nature, which are almost ambiguous to which kind they should be referred? But there is a mere and deep silence touching the nature and operation of those common adjuncts of things, as in nature: and only a resuming and repeating of the force and use of them in speech or argument. Therefore, because in a writing of this nature, I avoid all subtlety, my meaning touching this original or universal philosophy is thus, in a plain and gross description by negative: *That it be a receptacle for all such profitable observations and axioms as fall not within the compass of any of the special parts of philosophy or sciences, but are more common and of a higher stage.*

3.   Now that there are many of that kind need not be doubted. For example: is not the rule, *Si inaequalibus aequalia addas, omnia erunt inaequalia,* an axiom as well of justice as of the mathematics? and is there not a true coincidence between commutative and distributive justice, and arithmetical and geometrical proportion? Is not that other rule, *Quae in eodem tertio conveniunt, et inter se conveniunt,* a rule taken from the mathematics, but so potent in logic as all syllogisms are built upon it? Is not the observation, *Omnia mutantur, nil interit,*[60] a contemplation in philosophy thus, that the *quantum* of nature is eternal? in natural theology thus, that it requireth the same Omnipotency to make somewhat nothing, which at the first made nothing somewhat? accord-

ing to the Scripture, *Didici quod omnia opera, quae fecit Deus, perserverent in perpetuum; non possumus eis quicquam addere nec auferre.*[61] Is not the ground, which Machiavel wisely and largely discourseth concerning governments, that the way to establish and preserve them, is to reduce them *ad principia,* a rule in religion and nature, as well as in civil administration?[62] Was not the Persian magic a reduction or correspondence of the principles and architectures of nature to the rules and policy of governments? Is not the precept of a musician, to fall from a discord or harsh accord upon a concord or sweet accord, alike true in affection? Is not the trope of music, to avoid or slide from the close or cadence, common with the trope of rhetoric of deceiving expectation?[63] Is not the delight of the quavering upon a stop in music the same with the playing of light upon the water?

Splendet tremulo sub lumine pontus.[64]

Are not the organs of the senses of one kind with the organs of reflection, the eye with a glass, the ear with a cave or strait, determined and bounded? Neither are these only similitudes, as men of narrow observation may conceive them to be, but the same footsteps of nature, treading or printing upon several subjects or matters. This science, therefore, as I understand it, I may justly report as deficient: for I see sometimes the profounder sort of wits in handling some particular argument will now and then draw a bucket of water out of this well for their present use; but the spring-head thereof seemeth to me not to have been visited; being of so excellent use both for the disclosing of nature and the abridgment of art.

VI 1. This science being therefore first placed as a common parent, like unto Berecynthia, which had so much heavenly issue, *Omnes Caelicolas, omnes supera alta tenentes,*[65] we may return to the former distribution of the three philosophies, *divine, natural,* and *human.* And as concerning *divine philosophy* or *natural theology,* it is that knowledge or rudiment of knowledge concerning God, which may be obtained by the contemplation of His creatures; which knowledge may be truly termed *divine* in respect of the object, and *natural* in respect of the light. The bounds of this knowledge are, that it sufficeth to convince atheism, but not to inform religion: and therefore there was never miracle wrought by God to convert an atheist, because the light of nature might have led him to confess a God: but miracles have been wrought to convert idolaters and the superstitious, because no light of nature extendeth to declare the will and true worship of God. For as all works do show forth the power and skill of the workman, and not his image, so it is of the works of God, which do show the omnipotency and wisdom of the Maker, but not His image: and therefore therein the heathen opinion differeth from the sacred truth; for they supposed the world to be the image of God, and man to be an extract or compendious image of the world,[66] but the Scriptures never vouchsafe to attribute to the world that honour, as to be the image of God, but only *the work of His hands:*[67] neither do they speak of any other image of God, but man: wherefore by the contemplation of nature to induce and enforce the acknowledgment of God, and to demonstrate His power, providence, and goodness, is an excellent argument, and hath been excellently handled by divers. But on the other side, out of the contemplation of nature, or ground of human knowledges, to induce any verity or persuasion concerning the points of faith, is in my judgment not safe: *Da fidei quae fidei sunt.*[68] For the heathen themselves conclude as much in that excellent and divine fable of the golden chain: *That men and gods were not able to draw Jupiter down to the earth; but contrariwise Jupiter was able to draw them up to heaven.*[69] So as we ought not to attempt to draw down or submit the mysteries of God to our reason; but contrariwise to raise and advance our reason to the divine truth. So as in this part of knowledge, touching divine philosophy, I am so far from noting any deficience, as I rather note an excess: whereunto I have digressed because of the extreme prejudice which both religion and philosophy hath received and may receive, by being commixed together; as that which undoubtedly will make an heretical religion, and an imaginary and fabulous philosophy.

2.   Otherwise it is of the nature of angels and spirits, which is an appendix of theology both divine and natural, and is neither inscrutable nor interdicted; for although the Scripture saith, *Let no man deceive you in sublime discourse touching the worship of angels, pressing into*

that he knoweth not, etc.,[70] yet notwithstanding if you observe well that precept, it may appear thereby that there be two things only forbidden, adoration of them, and opinion fantastical of them, either to extol them further than appertaineth to the degree of a creature, or to extol a man's knowledge of them further than he hath ground. But the sober and grounded inquiry, which may arise out of the passages of holy Scriptures, or out of the gradations of nature, is not restrained. So of degenerate and revolted spirits, the conversing with them or the employment of them is prohibited, much more any veneration towards them; but the contemplation or science of their nature, their power, their illusions, either by Scripture or reason, is a part of spiritual wisdom. For so the apostle saith, *We are not ignorant of his stratagems.*[71] And it is no more unlawful to inquire the nature of evil spirits, than to inquire the force of poisons in nature, or the nature of sin and vice in morality. But this part touching angels and spirits I cannot note as deficient, for many have occupied themselves in it;[72] I may rather challenge it, in many of the writers thereof, as fabulous and fantastical.

VII   1.   Leaving therefore divine philosophy or natural theology (not Divinity or inspired theology, which we reserve for the last of all as the haven and sabbath of all man's contemplations), we will now proceed to natural philosophy. If then it be true that Democritus said, *That the truth of nature lieth hid in certain deep mines and caves,*[73] and if it be true likewise that the alchemists do so much inculcate, that Vulcan is a second nature, and imitateth that dexterously and compendiously, which nature worketh by ambages and length of time, it were good to divide natural philosophy into the mine and the furnace: and to make two professions or occupations of natural philosophers, some to be pioneers and some smiths; some to dig, and some to refine and hammer: and surely I do best allow of a division of that kind, though in more familiar and scholastical terms; namely, that these be the two parts of natural philosophy, the *inquisition of causes,* and the *production of effects; speculative,* and *operative; natural science,* and *natural prudence.* For as in civil matters there is a wisdom of discourse and a wisdom of direction; so is it in natural. And here I will make a request, that for the latter, or at least for a part thereof, I may revive and reintegrate the misapplied and abused name of *natural magic;*[74] which, in the true sense, is but *natural wisdom,* or *natural prudence;* taken according to the ancient acception, purged from vanity and superstition. Now although it be true, and I know it well, that there is an intercourse between causes and effects, so as both these knowledges, speculative and operative, have a great connection between themselves; yet because all true and fruitful natural philosophy hath a double scale or ladder, ascendent and descendent; ascending from experiments to the invention of causes, and descending from causes to the invention of new experiments; therefore I judge it most requisite that these two parts be severally considered and handled.

2.   *Natural science* or *theory* is divided into *physic* and *metaphysic:* wherein I desire it may be conceived that I use the word metaphysic in a differing sense from that that is received: and in like manner, I doubt not but it will easily appear to men of judgment, that in this and other particulars, wheresoever my conception and notion may differ from the ancient, yet I am studious to keep the ancient terms. For hoping well to deliver myself from mistaking, by the order and perspicuous expressing of that I do propound, I am otherwise zealous and affectionate to recede as little from antiquity, either in terms or opinions, as may stand with truth and the proficience of knowledge. And herein I cannot a little marvel at the philosopher Aristotle, that did proceed in such a spirit of difference and contradiction towards all antiquity: undertaking not only to frame new words of science at pleasure, but to confound and extinguish all ancient wisdom: insomuch as he never nameth or mentioneth an ancient author or opinion, but to confute and reprove;[75] wherein for glory, and drawing followers and disciples, he took the right course. For certainly there cometh to pass and hath place in human truth, that which was noted and pronounced in the highest truth: *Veni in nomine Patris, nec recipitis me; si quis venerit in nomine suo eum recipietis.*[76] But in this divine aphorism (considering to whom it was applied, namely to Antichrist, the highest deceiver) we may discern well that the coming in a man's own name, without regard of antiquity or paternity, is no good sign of truth, although it be joined with the fortune and success of an *Eum recipietis.* But for this

excellent person Aristotle, I will think of him that he learned that humour of his scholar, with whom it seemeth he did emulate, the one to conquer all opinions, as the other to conquer all nations; wherein nevertheless, it may be, he may at some men's hands that are of a bitter disposition get a like title as his scholar did:

> Felix terrarum praedo, non utile mundo
> Editus exemplum, etc.

So

> Felix doctrinae praedo.[77]

But to me, on the other side, that do desire as much as lieth in my pen to ground a sociable intercourse between antiquity and proficience, it seemeth best to keep way with antiquity *usque ad aras;* and therefore to retain the ancient terms, though I sometimes alter the uses and definitions, according to the moderate proceeding in civil government; where although there be some alteration, yet that holdeth which Tacitus wisely noteth, *Eadem Magistratuum vocabula.*[78]

3.   To return therefore to the use and acceptation of the term Metaphysic, as I do now understand the word; it appeareth, by that which hath been already said, that I intend *philosophia prima,* Summary Philosophy and Metaphysic, which heretofore have been confounded as one, to be two distinct things. For the one I have made as a parent or common ancestor to all knowledge; and the other I have now brought in as a branch or descendant of natural science. It appeareth likewise that I have assigned to Summary Philosophy the common principles and axioms which are promiscuous and indifferent to several sciences: I have assigned unto it likewise the inquiry touching the operation of the relative and adventive characters of essences, as *quantity, similitude, diversity, possibility,* and the rest: with this distinction and provision; that they be handled as they have efficacy in nature, and not logically. It appeareth likewise that Natural Theology, which heretofore hath been handled confusedly with Metaphysic, I have inclosed and bounded by itself. It is therefore now a question what is left remaining for Metaphysic; wherein I may without prejudice preserve thus much of the conceit of antiquity, that Physic should contemplate that which

is inherent in matter, and therefore transitory; and Metaphysic that which is abstracted and fixed. And again, that Physic should handle that which supposeth in nature only a being and moving; and Metaphysic should handle that which supposeth further in nature a reason, understanding, and platform. But the difference, perspicuously expressed, is most familiar and sensible. For as we divided natural philosophy in general into the *inquiry* of *causes,* and *productions* of *effects:* so that part which concerneth the inquiry of causes we do subdivide according to the received and sound division of causes; the one part, which is Physic, inquireth and handleth the *material* and *efficient causes;* and the other, which is Metaphysic, handleth the *formal* and *final causes.*[79]

4.   Physic, taking it according to the derivation, and not according to our idiom for *medicine,* is situate in a middle term or distance between Natural History and Metaphysic. For natural history describeth the variety of things; physic the causes, but variable or respective causes; and metaphysic the fixed and constant causes.

> Limus ut hic durescit, et haec ut cera liquescit,
> Uno eodemque igni:[80]

Fire is the cause of induration, but respective to clay; fire is the cause of colliquation, but respective to wax; but fire is no constant cause either of induration or colliquation: so then the physical causes are but the efficient and the matter. Physic hath three parts; whereof two respect nature united or collected, the third contemplateth nature diffused or distributed. Nature is collected either into one entire total, or else into the same principles or seeds. So as the first doctrine is touching the contexture or configuration of things, as *de mundo, de universitate rerum.* The second is the doctrine concerning the principles or originals of things. The third is the doctrine concerning all variety and particularity of things; whether it be of the differing substances, or their differing qualities and natures; whereof there needeth no enumeration, this part being but as a gloss, or paraphrase, that attendeth upon the text of natural history. Of these three I cannot report any as deficient. In what truth or perfection they are handled, I make not now any judgment; but they are parts of knowledge not deserted by the labour of man.

5. For Metaphysic, we have assigned unto it the inquiry of formal and final causes; which assignation, as to the former of them, may seem to be nugatory and void, because of the received and inveterate opinion that the inquisition of man is not competent to find out essential Forms or true differences: of which opinion we will take this hold, that the invention of Forms is of all other parts of knowledge the worthiest to be sought, if it be possible to be found.[81] As for the possibility, they are ill discoverers that think there is no land, when they can see nothing but sea. But it is manifest that Plato, in his opinion of Ideas, as one that had a wit of elevation situate as upon a cliff, did descry, *that Forms were the true object of knowledge;*[82] but lost the real fruit of his opinion, by considering of Forms as absolutely abstracted from matter, and not confined and determined by matter; and so turning his opinion upon theology, wherewith all his natural philosophy is infected.[83] But if any man shall keep a continual watchful and severe eye upon action, operation, and the use of knowledge, he may advise and take notice what are the Forms, the disclosures whereof are fruitful and important to the state of man. For as to the forms of substances, man only except, of whom it is said, *Formavit hominem de limo terrae, et spiravit in faciem ejus spiraculum vitae,* and not as of all other creatures, *Producant aquae, producat terra;*[84] the Forms of substances, I say, as they are now by compounding and transplanting multiplied, are so perplexed, as they are not to be inquired; no more than it were either possible or to purpose to seek in gross the Forms of those sounds which make words, which by composition and transposition of letters are infinite. But, on the other side, to inquire the Form of those sounds or voices which make simple letters is easily comprehensible; and being known, induceth and manifesteth the Forms of all words, which consist and are compounded of them. In the same manner to inquire the Form of a lion, of an oak, of gold; nay, of water, of air, is a vain pursuit: but to inquire the forms of sense, of voluntary motion, of vegetation, of colours, of gravity and levity, of density, of tenuity, of heat, of cold, and all other natures and qualities, which, like an alphabet, are not many, and of which the essences, upheld by matter, of all creatures do consist; to inquire, I say, the true Forms of these, is that part of metaphysic which we now define of. Not but that Physic doth make inquiry,

and take consideration of the same natures: but how? Only as to the *material and efficient causes* of them, and not as to the Forms. For example, if the cause of whiteness in snow or froth be inquired, and it be rendered thus, that the subtle intermixture of air and water is the cause, it is well rendered; but nevertheless is this the form of whiteness? No; but it is the efficient, which is ever but *vehiculum formae*.[85] This part of Metaphysic I do not find laboured and performed: whereat I marvel not; because I hold it not possible to be invented by that course of invention which hath been used; in regard that men, which is the root of all error, have made too untimely a departure and too remote a recess from particulars.

6.   But the use of this part of Metaphysic, which I report as deficient, is of the rest the most excellent in two respects: the one, because it is the duty and virtue of all knowledge to abridge the infinity of individual experience, as much as the conception of truth will permit, and to remedy the complaint of *vita brevis, ars longa;*[86] which is performed by uniting the notions and conceptions of sciences: for knowledges are as pyramids, whereof history is the basis. So of natural philosophy, the basis is natural history; the stage next the basis is physic; the stage next the vertical point is metaphysic. As for the vertical point, *opus quod operatur Deus a principio usque ad finem,*[87] the summary law of nature, we know not whether man's inquiry can attain unto it. But these three be the true stages of knowledge, and are to them that are depraved no better than the giant's hills:

> Ter sunt conati imponere Pelio Ossam,
> Scilicet atque Ossae frondosum involvere Olympum.[88]

But to those which refer all things to the glory of God, they are as the three acclamations, *Sancte, sancte, sancte!* holy in the description or dilatation of His works; holy in the connection or concatenation of them: and holy in the union of them in a perpetual and uniform law. And therefore the speculation was excellent in Parmenides and Plato, although but a speculation in them, that all things by scale did ascend to unity.[89] So then always that knowledge is worthiest which is charged with least multiplicity; which appeareth to be metaphysic; as that which considereth the simple Forms or differences of things, which are few in number, and the degrees and co-ordinations whereof make all this variety. The second respect, which valueth and commendeth this part of metaphysic, is that it doth enfranchise the power of man unto the greatest liberty and possibility of works and effects. For physic carrieth men in narrow and restrained ways, subject to many accidents of impediments, imitating the ordinary flexuous courses of nature; but *latae undique sunt sapientibus viae:*[90] to sapience, which was anciently defined to be *rerum divinarum et humanarum scientia,*[91] there is ever choice of means. For physical causes give light to new invention in *simili materia;* but whosoever knoweth any Form, knoweth the utmost possibility of superinducing that nature upon any variety of matter; and so is less restrained in operation, either to the basis of the matter, or the condition of the efficient; which kind of knowledge Salomon likewise, though in a more divine sense, elegantly describeth: *non arctabuntur gressus tui, et currens non habebis offendiculum.*[92] The ways of sapience are not much liable either to particularity or chance.

7.   The second part of metaphysic is the *inquiry of final causes,* which I am moved to report not as omitted, but as misplaced; and yet if it were but a fault in order, I would not speak of it: for order is matter of illustration, but pertaineth not to the substance of sciences. But this misplacing hath caused a deficience, or at least a great improficience in the sciences themselves. For the handling of final causes, mixed with the rest in physical inquiries, hath intercepted the severe and diligent inquiry of all real and physical causes, and given men the occasion to stay upon these satisfactory and specious causes, to the great arrest and prejudice of further discovery. For this I find done not only by Plato, who ever anchoreth upon that shore, but by Aristotle, Galen, and others which do usually likewise fall upon these flats of discoursing causes.[93] For to say that *the hairs of the eyelids are for a quickset and fence about the sight;* or that *the firmness of the skins and hides of living creatures is to defend them from the extremities of heat or cold;* or that *the bones are for the columns or beams, whereupon the frames of the bodies of living creatures are built:* or that *the leaves of trees are for protecting of the fruit;* or that *the clouds are for watering of the earth;* or that the *solidness of the earth is for the station and mansion of living creatures* and the like, is well inquired and collected in metaphysic, but in physic

they are impertinent. Nay, they are indeed but *remoraes* and hindrances to stay and slug the ship from further sailing; and have brought this to pass, that the search of the physical causes hath been neglected, and passed in silence. And therefore the natural philosophy of Democritus and some others (who did not suppose a mind or reason in the frame of things, but attributed the form thereof able to maintain itself to infinite essays or proofs of nature, which they term *fortune*) seemeth to me, as far as I can judge by the recital and fragments which remain unto us, in particularities of physical causes, more real and better inquired than that of Aristotle and Plato; whereof both intermingled final causes, the one as a part of theology, and the other as a part of logic, which were the favourite studies respectively of both those persons. Not because those final causes are not true, and worthy to be inquired, being kept within their own province; but because their excursions into the limits of physical causes hath bred a vastness and solitude in that tract. For otherwise, keeping their precincts and borders, men are extremely deceived if they think there is an enmity or repugnancy at all between them. For the cause rendered, that *the hairs about the eyelids are for the safeguard of the sight,* doth not impugn the cause rendered, that *pilosity is incident to orifices of moisture; muscosi fontes,*[94] etc. Nor the cause rendered, that *the firmness of hides is for the armour of the body against extremities of heat or cold,* doth not impugn the cause rendered, that *contraction of pores is incident to the outwardest parts, in regard of their adjacence to foreign or unlike bodies:* and so of the rest: both causes being true and compatible, the one declaring an *intention,* the other a *consequence* only. Neither doth this call in question, or derogate from Divine Providence, but highly confirm and exalt it. For as in civil actions he is the greater and deeper politique, that can make other men the instruments of his will and ends, and yet never acquaint them with his purpose, so as they shall do it and yet not know what they do, than he that imparteth his meaning to those he employeth; so is the wisdom of God more admirable, when nature intendeth one thing, and Providence draweth forth another, than if He had communicated to particular creatures and motions the characters and impressions of His Providence. And thus much for metaphysic: the latter part whereof I allow as extant, but with it confined to his proper place.

VIII

1. Nevertheless there remaineth yet another part of Natural Philosophy, which is commonly made a principal part and holdeth rank with Physic special and Metaphysic, which is Mathematic; but I think it more agreeable to the nature of things and to the light of order to place it as a branch of Metaphysic: for the subject of it being *quantity* (not *quantity indefinite,* which is but a *relative,* and belongeth to *philosophia prima,* as hath been said, but *quantity determined or proportionable*), it appeareth to be one of the essential Forms of things; as that that is causative in nature of a number of effects; insomuch as we see, in the schools both of Democritus and of Pythagoras,[95] that the one did ascribe figure to the first seeds of things, and the other did suppose numbers to be the principles and originals of things: and it is true also that of all other Forms, as we understand Forms, it is the most abstracted and separable from matter, and therefore most proper to Metaphysic; which hath likewise been the cause why it hath been better laboured and inquired than any of the other Forms, which are more immersed in matter. For it being the nature of the mind of man, to the extreme prejudice of knowledge, to delight in the spacious liberty of generalities, as in a champain region, and not in the inclosures of particularity; the Mathematics of all other knowledge were the goodliest fields to satisfy that appetite. But for the placing of this science, it is not much material: only we have endeavoured in these our partitions to observe a kind of perspective, that one part may cast light upon another.

2. The Mathematics are either *pure* or *mixed.* To the Pure Mathematics are those sciences belonging which handle *quantity determinate,* merely severed from any axioms of natural philosophy; and these are two, Geometry and Arithmetic; the one handling quantity continued, and the other dissevered. Mixed hath for subject some axioms or parts of natural philosophy, and considereth *quantity determined,* as it is auxiliary and incident unto them. For many parts of nature can neither be invented with sufficient subtlety, nor demonstrated with sufficient perspicuity, nor accommodated unto use with sufficient dexterity, without the aid and intervening of the mathematics; of which sort are *perspective, music, astronomy, cosmography, architecture, enginery,* and divers others. In the Mathematics I can report no deficience, except it

be that men do not sufficiently understand the excellent use of the Pure Mathematics, in that they do remedy and cure many defects in the wit and faculties intellectual. For if the wit be too dull, they sharpen it; if too wandering, they fix it; if too inherent in the sense, they abstract it. So that as tennis is a game of no use in itself, but of great use in respect it maketh a quick eye and a body ready to put itself into all postures; so in the Mathematics, that use which is collateral and intervenient is no less worthy than that which is principal and intended. And as for the Mixed Mathematics, I may only make this prediction, that there cannot fail to be more kinds of them, as nature grows further disclosed. Thus much of Natural Science, or the part of nature speculative.

3.   For Natural Prudence, or the part operative of Natural Philosophy, we will divide it into three parts, experimental, philosophical, and magical; which three parts active have a correspondence and analogy with the three parts speculative, natural history, physic, and metaphysic: for many operations have been invented, sometimes by a casual incidence and occurrence, sometimes by a purposed experiment: and of those which have been found by an intentional experiment, some have been found out by varying or extending the same experiments, some by transferring and compounding divers experiments the one into the other, which kind of invention an empiric may manage. Again, by the knowledge of physical causes there cannot fail to follow many indications and designations of new particulars, if men in their speculation will keep one eye upon use and practice. But these are but coastings along the shore, *Premendo littus iniquum:*[96] for it seemeth to me there can hardly be discovered any radical or fundamental alterations and innovations in nature, either by the fortune and essays of experiments, or by the light and direction of physical causes.

If therefore we have reported Metaphysic deficient, it must follow that we do the like of Natural Magic, which hath relation thereunto. For as for the Natural Magic whereof now there is mention in books, containing certain credulous and superstitious conceits and observations of sympathies and antipathies, and hidden properties, and some frivolous experiments, strange rather by disguisement than in themselves; it is as far differing in truth of nature from such a knowledge as we require, as the story of King Arthur of Britain, or Hugh of Bordeaux, differs from Caesar's Commentaries in truth of story. For it is manifest that Caesar did greater things *de vero* than those imaginary heroes were feigned to do; but he did them not in that fabulous manner. Of this kind of learning the fable of Ixion[97] was a figure, who designed to enjoy Juno, the goddess of power; and instead of her had copulation with a cloud, of which mixture were begotten centaurs and chimeras. So whosoever shall entertain high and vaporous imaginations, instead of a laborious and sober inquiry of truth, shall beget hopes and beliefs of strange and impossible shapes. And therefore we may note in these sciences which hold so much of imagination and belief, as this degenerate Natural Magic, Alchemy, Astrology, and the like, that in their propositions the description of the means is ever more monstrous than the pretence or end. For it is a thing more probable, that he that knoweth well the natures of *weight,* of *colour,* of *pliant* and *fragile* in respect of the hammer, of *volatile* and *fixed* in respect of the fire, and the rest, may superinduce upon some metal the nature and Form of gold by such mechanique as longeth to the production of the natures afore rehearsed, than that some grains of the medicine projected should in a few moments of time turn a sea of quicksilver or other material into gold: so it is more probable that he that knoweth the nature of arefaction, the nature of assimilation of nourishment to the thing nourished, the manner of increase and clearing of spirits, the manner of the depredations which spirits make upon the humours and solid parts, shall by ambages of diets, bathings, anointings, medicines, motions, and the like, prolong life, or restore some degree of youth or vivacity, than that it can be done with the use of a few drops or scruples of a liquor or receipt.

To conclude, therefore, the true Natural Magic, which is that great liberty and latitude of operation which dependeth upon the knowledge of Forms, I may report deficient, as the relative thereof is. To which part, if we be serious, and incline not to vanities and plausible discourse, besides the deriving and deducing the operations themselves from Metaphysic, there are pertinent two points of much purpose, the one by way of preparation, the other by way of caution: the first is, that there be made a kalendar, resembling an inventory of the estate of man, containing all the inventions, being the works or fruits of nature or art, which are now extant, and whereof man is already possessed; out of which doth naturally result a note, what things are yet held impossible, or not invented: which kalendar will be the more artificial and serviceable, if to every reputed impossibility you add what thing is extant which cometh the nearest in degree to that impossibility; to the end that by these optatives and potentials man's inquiry may be the more awake in deducing direction of works from the speculation of causes: and secondly, that those experiments be not only esteemed which have an immediate and present use, but those principally which are of most universal consequence for invention of other experiments, and those which give most light to the invention of causes; for the invention of the mariner's needle, which giveth the direction, is of no less benefit for navigation than the invention of the sails which give the motion.

4.   Thus have I passed through Natural Philosophy, and the deficiencies thereof; wherein if I have differed from the ancient and received doctrines, and thereby shall move contradiction; for my part, as I affect not to dissent, so I purpose not to contend. If it be truth,

Non canimus surdis, respondent omnia sylvae;[98]

the voice of nature will consent, whether the voice of man do or no. And as Alexander Borgia was wont to say of the expedition of the French for Naples, that they came with chalk in their hands to mark up their lodgings, and not with weapons to fight; so I like better that entry of truth which cometh peaceably, with chalk to mark up those minds which are capable to lodge and harbour it, than that which cometh with pugnacity and contention.[99]

5.   But there remaineth a division of natural philosophy according to the report of the inquiry, and nothing concerning the matter or subject; and that is positive and considerative; when the inquiry reporteth either an assertion or a doubt. These doubts or *non liquets* are of two sorts, particular and total. For the first, we see a good example thereof in Aristotle's Problems, which deserved to have had a better continuance; but so nevertheless as there is one point whereof warning is to be given and taken. The registering of doubts hath two excellent uses: the one, that it saveth philosophy from errors and falsehoods; when that which is not fully appearing is not collected into assertion, whereby error

might draw error, but reserved in doubt: the other, that the entry of doubts are as so many suckers or sponges to draw use of knowledge; insomuch as that which, if doubts had not preceded, a man should never have advised, but passed it over without note, by the suggestion and solicitation of doubts, is made to be attended and applied. But both these commodities do scarcely countervail an inconvenience, which will intrude itself if it be not debarred; which is, that when a doubt is once received, men labour rather how to keep it a doubt still, than how to solve it; and accordingly bend their wits. Of this we see the familiar example in lawyers and scholars, both which, if they have once admitted a doubt, it goeth ever after authorised for a doubt. But that use of wit and knowledge is to be allowed, which laboureth to make doubtful things certain, and not those which labour to make certain things doubtful. Therefore these kalendars of doubts I commend as excellent things; so that there be this caution used, that when they be thoroughly sifted and brought to resolution, they be from thenceforth omitted, decarded, and not continued to cherish and encourage men in doubting. To which kalendar of doubts or problems, I advise be annexed another kalendar, as much or more material, which is a Kalendar of popular errors: I mean chiefly in natural history, such as pass in speech and conceit, and are nevertheless apparently detected and convicted of untruth: that man's knowledge be not weakened nor imbased by such dross and vanity.

As for the doubts or *non liquets* general or in total, I understand those differences of opinions touching the principles of nature, and the fundamental points of the same, which have caused the diversity of sects, schools, and philosophies, as that of Empedocles, Pythagoras, Democritus, Parmenides, and the rest. For although Aristotle, as though he had been of the race of the Ottomans, thought he could not reign except the first thing he did he killed all his brethren;[100] yet to those that seek Truth and not magistrality, it cannot but seem a matter of great profit, to see before them the several opinions touching the foundations of nature: not for any exact truth that can be expected in those theories; for as the same phenomena in astronomy are satisfied by the received astronomy of the diurnal motion, and the proper motions of the planets, with their eccentrics and epicycles, and likewise by the theory of Copernicus,[101] who supposed the earth to move (and the calculations are indifferently agreeable to both), so the ordinary face and view of experience is many times satisfied by several theories and philosophies; whereas to find the real truth requireth another manner of severity and attention. For as Aristotle saith,[102] that children at the first will call every woman mother, but afterward they come to distinguish according to truth; so experience, if it be in childhood, will call every philosophy mother, but when it cometh to ripeness, it will discern the true mother. So as in the mean time it is good to see the several glosses and opinions upon nature, whereof it may be every one in some one point hath seen clearer than his fellows: therefore I wish some collection to be made, painfully and understandingly, *de antiquis philosophiis*, out of all the possible light which remaineth to us of them: which kind of work I find deficient. But here I must give warning, that it be done distinctly and severedly; the philosophies of every one throughout by themselves; and not by titles packed and fagotted up together, as hath been done by Plutarch. For it is the harmony of a philosophy in itself which giveth it light and credence; whereas if it be singled and broken, it will seem more foreign and dissonant. For as when I read in Tacitus the actions of Nero, or Claudius, with circumstances of times, inducements, and occasions, I find them not so strange; but when I read them in Suetonius Tranquillus, gathered into titles and bundles, and not in order of time, they seem more monstrous and incredible: so is it of any philosophy reported entire, and dismembered by articles. Neither do I exclude opinions of latter times to be likewise represented in this kalendar of sects of philosophy, as that of Theophrastus Paracelsus,[103] eloquently reduced into a harmony by the pen of Severinus the Dane:[104] and that of Telesius[105] and his scholar Donius, being as a pastoral philosophy, full of sense, but of no great depth; and that of Fracastorius,[106] who, though he pretended not to make any new philosophy, yet did use the absoluteness of his own sense upon the old; and that of Gilbertus our countryman,[107] who revived, with some alterations and demonstrations, the opinions of Xenophanes: and any other worthy to be admitted.

6.   Thus have we now dealt with two of the three beams of man's knowledge; that is, *radius directus,* which is referred to nature; *radius refractus,* which is referred to God, and cannot report truly because of

the inequality of the *medium*. There resteth *radius reflexus,* whereby man beholdeth and contemplateth himself.

IX 1. We come therefore now to that knowledge whereunto the ancient oracle directeth us, which is *the knowledge of ourselves;*[108] which deserveth the more accurate handling, by how much it toucheth us more nearly. This knowledge, as it is the end and term of natural philosophy in the intention of man, so notwithstanding it is but a portion of natural philosophy in the continent of nature: and generally let this be a rule, that all partitions of knowledges be accepted rather for lines and veins than for sections and separations; and that the continuance and entireness of knowledge be preserved. For the contrary hereof hath made particular sciences to become barren, shallow, and erroneous, while they have not been nourished and maintained from the common fountain. So we see Cicero the orator complained of Socrates and his school that he was the first that separated philosophy and rhetoric;[109] whereupon rhetoric became an empty and verbal art. So we may see that the opinion of Copernicus touching the rotation of the earth, which astronomy itself cannot correct, because it is not repugnant to any of the phenomena, yet natural philosophy may correct. So we see also that the science of medicine, if it be destituted and forsaken by natural philosophy, it is not much better than an empirical practice. With this reservation therefore we proceed to human philosophy or humanity, which hath two parts: the one considereth man *segregate* or *distributively;* the other *congregate* or *in society.* So as human philosophy is either simple and particular, or conjugate and civil. Humanity particular consisteth of the same parts whereof man consisteth; that is, of knowledges which respect the body, and of knowledges that respect the mind. But before we distribute so far, it is good to constitute. For I do take the consideration in general and at large of human nature to be fit to be emancipate and made a knowledge by itself: not so much in regard of those delightful and elegant discourses which have been made of the dignity of man, of his miseries, of his state and life, and the like adjuncts of his common and undivided nature; but chiefly in regard of the knowledge concerning the sympathies and concordances between the mind and body, which being mixed cannot be properly assigned to the sciences of either.

2.   This knowledge hath two branches: for as all leagues and amities consist of mutual intelligence and mutual offices, so this league of mind and body hath these two parts; how the one discloseth the other, and how the one worketh upon the other; discovery and impression. The former of these hath begotten two arts, both of *prediction* or *prenotion;* whereof the one is honoured with the inquiry of Aristotle, and the other of Hippocrates.[110] And although they have of later time been used to be coupled with superstitious and fantastical arts, yet being purged and restored to their true state, they have both of them a solid ground in nature, and a profitable use in life. The first is *physiognomy,* which discovereth the disposition of the mind by the lineaments of the body: the second is the *exposition of natural dreams,* which discovereth the state of the body by the imaginations of the mind. In the former of these I note a deficience. For Aristotle hath very ingeniously and diligently handled the factures of the body, but not the gestures of the body, which are no less comprehensible by art, and of greater use and advantage.[111] For the lineaments of the body do disclose the disposition and inclination of the mind in general; but the motions of the countenance and parts do not only so, but do further disclose the present humour and state of the mind and will. For as your Majesty saith most aptly and elegantly, *As the tongue speaketh to the ear so the gesture speaketh to the eye.*[112] And therefore a number of subtle persons, whose eyes do dwell upon the faces and fashions of men, do well know the advantage of this observation, as being most part of their ability; neither can it be denied, but that it is a great discovery of dissimulations, and a great direction in business.

3.   The latter branch, touching *impression,* hath not been collected into art, but hath been handled dispersedly; and it hath the same relation or *antistrophe* that the former hath. For the consideration is double: either *how, and how far the humours and affects of the body do alter or work upon the mind; or again, how and how far the passions or apprehensions of the mind do alter or work upon the body.* The former of these hath been inquired and considered as a part and appendix of medicine, but much more as a part of religion or superstition. For the

physician prescribeth cures of the mind in phrensies and melancholy passions; and pretendeth also to exhibit medicines to exhilarate the mind, to confirm the courage, to clarify the wits, to corroborate the memory, and the like: but the scruples and superstitions of diet and other regimen of the body in the sect of the Pythagoreans, in the heresy of the Manichees, and in the law of Mahomet, do exceed. So likewise the ordinances in the ceremonial law, interdicting the eating of the blood and the fat, distinguishing between beasts clean and unclean for meat, are many and strict. Nay the faith itself being clear and serene from all clouds of ceremony, yet retaineth the use of fastings, abstinences, and other macerations and humiliations of the body, as things real, and not figurative. The root and life of all of which prescripts is, besides the ceremony, the consideration of that dependency which the affections of the mind are submitted unto upon the state and disposition of the body. And if any man of weak judgment do conceive that this suffering of the mind from the body doth either question the immortality, or derogate from the sovereignty of the soul, he may be taught in easy instances that the infant in the mother's womb is compatible with the mother and yet separable;[113] and the most absolute monarch is sometimes led by his servants and yet without subjection. As for the reciprocal knowledge, which is the operation of the conceits and passions of the mind upon the body, we see all wise physicians, in the prescriptions of their regiments to their patients, do ever consider *accidentia animi* as of great force to further or hinder remedies or recoveries: and more especially it is an inquiry of great depth and worth concerning imagination, how and how far it altereth the body proper of the imaginant. For although it hath a manifest power to hurt, it followeth not it hath the same degree of power to help; no more than a man can conclude, that because there be pestilent airs able suddenly to kill a man in health, therefore there should be sovereign airs able suddenly to cure a man in sickness. But the inquisition of this part is of great use, though it needeth, as Socrates said, *a Delian diver,*[114] being difficult and profound. But unto all this knowledge *de communi vinculo,* of the concordances between the mind and the body, that part of inquiry is most necessary, which considereth of the seats and domiciles which the several faculties of the mind do take and occupy in the organs of the body; which knowledge hath been attempted, and is controverted, and deserveth to be much better inquired. For the opinion of Plato,[115] who placed the *understanding in the brain, animosity* (which he did unfitly call *anger,* having a greater mixture with *pride*) *in the heart,* and *concupiscence or sensuality* in the *liver,* deserveth not to be despised; but much less to be allowed. So then we have constituted, as in our own wish and advice, the inquiry touching human nature entire, as a just portion of knowledge to be handled apart.

X 1.   The knowledge that concerneth man's body is divided as the good of man's body is divided, unto which it referreth. The good of man's body is of four kinds, Health, Beauty, Strength, and Pleasure: so the knowledges are Medicine, or art of Cure; art of Decoration, which is called Cosmetic; art of Activity, which is called Athletic; and art Voluptuary, which Tacitus truly calleth *eruditus luxus.*[116] This subject of man's body is of all other things in nature most susceptible of remedy; but then that remedy is most susceptible of error. For the same subtlety of the subject doth cause large possibility and easy failing; and therefore the inquiry ought to be the more exact.

2.   To speak therefore of Medicine, and to resume that we have said, ascending a little higher: the ancient opinion that man was *microcosmus,* an abstract or model of the world, hath been fantastically strained by Paracelsus[117] and the alchemists, as if there were to be found in man's body certain correspondences and parallels, which should have respect to all varieties of things, as stars, planets, minerals, which are extant in the great world. But thus much is evidently true, that of all substances which nature hath produced, man's body is the most extremely compounded. For we see herbs and plants are nourished by earth and water; beasts for the most part by herbs and fruits; man by the flesh of beasts, birds, fishes, herbs, grains, fruits, water, and the manifold alterations, dressings, and preparations of the several bodies, before they come to be his food and aliment. Add hereunto, that beasts have a more simple order of life, and less change of affections to work upon their bodies: whereas man in his mansion, sleep, exercise, passions, hath infinite variations: and it cannot be denied but that the Body of man

of all other things is of the most compounded mass. The Soul on the other side is the simplest of substances, as is well expressed:

> Purumque reliquit
> Aethereum sensum atque aurai simplicis ignem.[118]

So that it is no marvel though the soul so placed enjoy no rest, if that principle be true, that *Motus rerum est rapidus extra locum, placidus in loco*. But to the purpose: this variable composition of man's body hath made it as an instrument easy to distemper; and therefore the poets did well to conjoin Music and Medicine in Apollo,[119] because the office of Medicine is but to tune this curious harp of man's body and to reduce it to harmony. So then the subject being so variable, hath made the art by consequence more conjectural; and the art being conjectural hath made so much the more place to be left for imposture. For almost all other arts and sciences are judged by acts, or masterpieces, as I may term them, and not by the successes and events. The lawyer is judged by the virtue of his pleading, and not by the issue of the cause; the master in the ship is judged by the directing his course aright, and not by the fortune of the voyage; but the physician, and perhaps the politique, hath no particular acts demonstrative of his ability, but is judged most by the event; which is ever but as it is taken: for who can tell if a patient die or recover, or if a state be preserved or ruined, whether it be art or accident? And therefore many times the impostor is prized, and the man of virtue taxed. Nay, we see the weakness and credulity of men is such, as they will often prefer a mountebank[120] or witch before a learned physician. And therefore the poets were clear-sighted in discerning this extreme folly, when they made Aesculapius and Circe brother and sister, both children of the sun, as in the verses,

> Ipse repertorem medicinae talis et artis
> Fulmine *Phoebigenam* Stygias detrusit ad undas:[121]

And again,

> Dives inaccessos ubi *Solis filia* lucos, etc.[122]

For in all times, in the opinion of the multitude, witches and old women and impostors have had a competition with physicians. And what followeth? Even this, that physicians say to themselves as Salomon express-eth it upon a higher occasion; *If it befall to me as befalleth to the fools, why should I labour to be more wise?*[123] And therefore I cannot much blame physicians, that they use commonly to intend some other art or practice, which they fancy more than their profession. For you shall have of them antiquaries, poets, humanists, statesmen, merchants, divines, and in every of these better seen than in their profession; and no doubt upon this ground, that they find that mediocrity and excel-lency in their art maketh no difference in profit or reputation towards their fortune; for the weakness of patients, and sweetness of life, and nature of hope, maketh men depend upon physicians with all their defects. But nevertheless, these things which we have spoken of, are courses begotten between a little occasion, and a great deal of sloth and default; for if we will excite and awake our observation, we shall see in familiar instances what a predominant faculty the *subtlety of spirit* hath over the *variety of matter or form:* nothing more variable than faces and countenances: yet men can bear in memory the infinite distinctions of them; nay, a painter with a few shells of colours, and the benefit of his eye and habit of his imagination, can imitate them all that ever have been, are, or may be, if they were brought before him: nothing more variable than voices; yet men can likewise discern them personally: nay, you shall have a buffoon or *pantomimus,*[124] will express as many as he pleaseth. Nothing more variable than the differing sounds of words; yet men have found the way to reduce them to a few simple letters. So that it is not the insufficiency or incapacity of man's mind, but it is the remote standing or placing thereof, that breedeth these mazes and incomprehensions: for as the sense afar off is full of mis-taking, but is exact at hand, so is it of the understanding; the remedy whereof is, not to quicken or strengthen the organ, but to go nearer to the object; and therefore there is no doubt but if the physicians will learn and use the true approaches and avenues of nature, they may as-sume as much as the poet saith:

> Et quoniam variant morbi, variabimus artes;
> Mille mali species, mille salutis erunt.[125]

Which that they should do, the nobleness of their art doth deserve;

well shadowed by the poets, in that they made Aesculapius to be the son of the sun, the one being the fountain of life, the other as the second stream: but infinitely more honoured by the example of our Saviour, who made the body of man the object of His miracles, as the soul was the object of His doctrine. For we read not that ever He vouchsafed to do any miracle about honour or money, except that one for giving tribute to Caesar;[126] but only about the preserving, sustaining, and healing the body of man.

3. Medicine is a science which hath been, as we said, more professed than laboured, and yet more laboured than advanced, the labour having been, in my judgment, rather in circle than in progression. For I find much iteration, but small addition. It considereth *causes of diseases, with the occasions or impulsions;* the *diseases themselves,* with the *accidents;* and the *cures,* with the *preservations.* The deficiencies which I think good to note, being a few of many, and those such as are of a more open and manifest nature, I will enumerate, and not place.

4. The first is the discontinuance of the ancient and serious diligence of Hippocrates,[127] which used to set down a narrative of the special cases of his patients, and how they proceeded, and how they were judged by recovery or death. Therefore having an example proper in the father of the art, I shall not need to allege an example foreign, of the wisdom of the lawyers, who are careful to report new cases and decisions for the direction of future judgments. This continuance of *medicinal history* I find deficient; which I understand neither to be so infinite as to extend to every common case, nor so reserved as to admit none but wonders: for many things are new in the manner, which are not new in the kind; and if men will intend to observe, they shall find much worthy to observe.

5. In the inquiry which is made by Anatomy, I find much deficience: for they inquire of the *parts,* and their *substances, figures,* and *collocations;* but they inquire not of the *diversities of the parts,* the *secrecies of the passages,* and the *seats or nestlings of the humours,* nor much of the *footsteps and impressions of diseases:* the reason of which omission I suppose to be, because the first inquiry may be satisfied in the view of one or a few anatomies: but the latter, being comparative and casual, must arise from the view of many. And as to the diversity of parts, there is no doubt but the facture or framing of the inward parts is as

full of difference as the outward, and in that is the *cause continent* of many diseases; which not being observed, they quarrel many times with the humours, which are not in fault; the fault being in the very frame and mechanique of the part, which cannot be removed by medicine alterative, but must be accommodate and palliate by diets and medicines familiar. As for the passages and pores, it is true which was anciently noted, that the more subtle of them appear not in anatomies, because they are shut and latent in dead bodies, though they be open and manifest in live: which being supposed, though the inhumanity of *anatomia vivorum* was by Celsus justly reproved;[128] yet in regard of the great use of this observation, the inquiry needed not by him so slightly to have been relinquished altogether, or referred to the casual practices of surgery; but might have been well diverted upon the dissection of beasts alive, which notwithstanding the dissimilitude of their parts, may sufficiently satisfy this inquiry. And for the humours, they are commonly passed over in anatomies as purgaments; whereas it is most necessary to observe, what cavities, nests, and receptacles the humours do find in the parts, with the differing kind of the humour so lodged and received. And as for the footsteps of diseases and their devastations of the inward parts, imposthumations, exulcerations, discontinuations, putrefactions, consumptions, contractions, extensions, convulsions, dislocations, obstructions, repletions, together with all preternatural substances, as stones, carnosities, excrescences, worms, and the like; they ought to have been exactly observed by multitude of anatomies, and the contribution of men's several experiences, and carefully set down, both historically, according to the appearances, and artificially, with a reference to the diseases and symptoms which resulted from them, in case where the anatomy is of a defunct patient; whereas now, upon opening of bodies, they are passed over slightly and in silence.

6.   In the inquiry of diseases, they do abandon the cures of many, some as in their nature incurable, and others as past the period of cure; so that Sulla and the Triumvirs never proscribed so many men to die, as they do by their ignorant edicts: whereof numbers do escape with less difficulty than they did in the Roman proscriptions. Therefore I will not doubt to note as a deficience, that they inquire not the perfect cures of many

diseases, or extremities of diseases; but pronouncing them incurable, do enact a law of neglect, and exempt ignorance from discredit.

7.   Nay, further, I esteem it the office of a physician not only to restore health, but to mitigate pain and dours; and not only when such mitigation may conduce to recovery, but when it may serve to make a fair and easy passage: for it is no small felicity which Augustus Caesar was wont to wish to himself, that same Euthanasia;[129] and which was especially noted in the death of Antoninus Pius, whose death was after the fashion and semblance of a kindly and pleasant sleep. So it is written of Epicurus, that after his disease was judged desperate, he drowned his stomach and senses with a large draught and ingurgitation of wine; whereupon the epigram was made, *Hinc Stygias ebrius hausit aquas;*[130] he was not sober enough to taste any bitterness of the Stygian water. But the physicians contrariwise do make a kind of scruple and religion to stay with the patient after the disease is deplored; whereas, in my judgment, they ought both to inquire the skill and to give the attendances for the facilitating and assuaging of the pains and agonies of death.

8.   In the consideration of the cures of diseases, I find a deficience in the receipts of propriety,[131] respecting the particular cures of diseases: for the physicians have frustrated the fruit of tradition and experience by their magistralities, in adding, and taking out, and changing *quid pro quo,* in their receipts at their pleasures; commanding so over the medicine, as the medicine cannot command over the disease: for except it be treacle and *mithridatum,*[132] and of late *diascordium,* and a few more, they tie themselves to no receipts severely and religiously: for as to the confections of sale which are in the shops, they are for readiness and not for propriety; for they are upon general intentions of purging, opening, comforting, altering, and not much appropriate to particular diseases: and this is the cause why empirics and old women are more happy many times in their cures than learned physicians, because they are more religious in holding their medicines. Therefore here is the deficience which I find, that physicians have not, partly out of their own practice, partly out of the constant probations reported in books, and partly out of the traditions of empirics, set down and delivered over certain experimental medicines for the cure of particular diseases, besides their own conjectural and magistral descriptions.

For as they were the men of the best composition in the state of Rome, which either being consuls inclined to the people, or being tribunes inclined to the senate; so in the matter we now handle, they be the best physicians, which being learned incline to the traditions of experience, or being empirics incline to the methods of learning.

9.   In preparation of medicines, I do find strange, especially considering how mineral medicines have been extolled,[133] and that they are safer for the outward than inward parts, that no man hath sought to make an imitation by art of natural baths and medicinable fountains: which nevertheless are confessed to receive their virtues from minerals: and not so only, but discerned and distinguished from what particular mineral they receive tincture, as sulphur, vitriol, steel, or the like; which nature, if it may be reduced to compositions of art, both the variety of them will be increased, and the temper of them will be more commanded.

10.   But lest I grow to be more particular than is agreeable either to my intention or to proportion, I will conclude this part with the note of one deficience more, which seemeth to me of greatest consequence; which is, that the prescripts in use are too compendious to attain their end: for, to my understanding, it is a vain and flattering opinion to think any medicine can be so sovereign or so happy, as that the receipt or use of it can work any great effect upon the body of man. It were a strange speech, which spoken, or spoken oft, should reclaim a man from a vice to which he were by nature subject: it is order, pursuit, sequence, and interchange of application, which is mighty in nature; which although it require more exact knowledge in prescribing, and more precise obedience in observing, yet is recompensed with the magnitude of effects. And although a man would think, by the daily visitations of the physicians, that there were a pursuance in the cure: yet let a man look into their prescripts and ministrations, and he shall find them but inconstancies and every day's devices, without any settled providence or project. Not that every scrupulous or superstitious prescript is effectual, no more than every straight way is the way to heaven; but the truth of the direction must precede severity of observance.[134]

11.   For Cosmetic, it hath parts civil, and parts effeminate: for cleanness of body was ever esteemed to proceed from a due reverence to God, to society, and to ourselves. As for artificial decoration, it is well

worthy of the deficiencies which it hath; being neither fine enough to deceive, nor handsome to use, nor wholesome to please.

12.   For Athletic, I take the subject of it largely, that is to say, for any point of ability whereunto the body of man may be brought, whether it be of *activity,* or of *patience;* whereof activity hath two parts, *strength and swiftness;* and patience likewise hath two parts, *hardness against wants and extremities,* and *endurance of pain or torment;* whereof we see the practices in tumblers, in savages, and in those that suffer punishment: nay, if there be any other faculty which falls not within any of the former divisions, as in those that dive, that obtain a strange power of containing respiration, and the like, I refer to it this part. Of these things the practices are known, but the philosophy that concerneth them is not much inquired; the rather, I think, because they are supposed to be obtained, either by an aptness of nature, which cannot be taught, or only by continual custom, which is soon prescribed: which though it be not true, yet I forbear to note any deficiencies: for the Olympian games are down long since, and the mediocrity of these things is for use; as for the excellency of them it serveth for the most part but for mercenary ostentation.

13.   For *arts of pleasure sensual,* the chief deficience in them is of laws to repress them.[135] For as it hath been well observed, that the arts which flourish in times while virtue is in growth, are *military;* and while virtue is in state, are *liberal;* and while virtue is in declination, are *voluptuary;* so I doubt that this age of the world is somewhat upon the descent of the wheel. With arts voluptuary I couple practices joculary; for the deceiving of the senses is one of the pleasures of the senses. As for games of recreation, I hold them to belong to civil life and education. And thus much of that particular human philosophy which concerns the body, which is but the tabernacle of the mind.

XI   1.   For Human Knowledge which concerns the Mind, it hath two parts; the one that inquireth of the substance or nature of the soul or mind, the other that inquireth of the faculties or functions thereof. Unto the first of these, the considerations of the original of the soul, whether it be native or adventive, and how far it is exempted from laws of matter, and of the immortality thereof, and many other points, do appertain: which have been not more laboriously inquired than variously reported; so as the travail therein taken seemeth to have been rather in a maze than in a way. But although I am of opinion that this knowledge may be more really and soundly inquired, even in nature, than it hath been; yet I hold that in the end it must be bounded by religion, or else it will be subject to deceit and delusion: for as the substance of the soul in the creation was not extracted out of the mass of heaven and earth by the benediction of a *producat* but was immediately inspired from God: so it is not possible that it should be (otherwise than by accident) subject to the laws of heaven and earth, which are the subject of philosophy; and therefore the true knowledge of the nature and state of the soul must come by the same inspiration that gave the substance. Unto this part of knowledge touching the soul there be two appendices; which, as they have been handled, have rather vapoured forth fables than kindled truth, Divination and Fascination.

2.   *Divination* hath been anciently and fitly divided into artificial and natural; whereof artificial is, when the mind maketh a prediction by argument, concluding upon signs and tokens; natural is when the mind hath a presention by an internal power, without the inducement of a sign. Artificial is of two sorts; either when the argument is coupled with a derivation of causes, which is rational; or when it is only grounded upon a coincidence of the effect, which is experimental: whereof the latter for the most part is superstitious; such as were the heathen observations upon the inspection of sacrifices, the flights of birds, the swarming of bees; and such as was the Chaldean astrology, and the like. For artificial divination, the several kinds thereof are distributed amongst particular knowledges. The astronomer hath his predictions, as of conjunctions, aspects, eclipses, and the like. The physician hath his predictions of death, of recovery, of the accidents and issues of diseases. The Politique hath his predictions; *O urbem venalem, et cito perituram, si emptorem invenerit!*[136] which stayed not long to be performed, in Sulla first, and after in Caesar. So as these predictions are now impertinent, and to be referred over. But the divination which springeth from the internal nature of the soul, is that which we now speak of; which hath been made to be of two sorts, primitive and by influxion. Primitive is

grounded upon the supposition, that the mind, when it is withdrawn and collected into itself, and not diffused into the organs of the body, hath some extent and latitude of prenotion; which therefore appeareth most in sleep, in ecstasies, and near death, and more rarely in waking apprehensions; and is induced and furthered by those abstinences and observances which make the mind most to consist in itself. By influxion, is grounded upon the conceit that the mind, as a mirror or glass, should take illumination from the foreknowledge of God and spirits:[137] unto which the same regiment doth likewise conduce. For the retiring of the mind within itself, is the state which is most susceptible of divine influxions; save that it is accompanied in this case with a fervency and elevation, which the ancients noted by *fury*, and not with a repose and quiet, as it is in the other.

3.   *Fascination* is the power and act of imagination intensive upon other bodies than the body of the imaginant, for of that we spake in the proper place: wherein the school of Paracelsus, and the disciples of pretended Natural Magic have been so intemperate, as they have exalted the power of the imagination to be much one with the power of miracle-working faith; others, that draw nearer to probability, calling to their view the secret passages of things, and specially of the contagion that passeth from body to body, do conceive it should likewise be agreeable to nature, that there should be some transmissions and operations from spirit to spirit without the mediation of the senses; whence the conceits have grown, now almost made civil, of the mastering spirit, and the force of confidence, and the like. Incident unto this is the inquiry how to raise and fortify the imagination: for if the imagination fortified have power, then it is material to know how to fortify and exalt it. And herein comes in crookedly and dangerously a palliation of a great part of Ceremonial Magic. For it may be pretended that Ceremonies, Characters, and Charms, do work, not by any tacit or sacramental contract with evil spirits, but serve only to strengthen the imagination of him that useth it: as images are said by the Roman church to fix the cogitations, and raise the devotions of them that pray before them. But for mine own judgment, if it be admitted that imagination hath power, and that Ceremonies fortify imagination, and that they be used sincerely and intentionally for that purpose;[138] yet I should

hold them unlawful, as opposing to that first edict which God gave unto man, *In sudore vultus comedes panem tuum.*[139] For they propound those noble effects, which God hath set forth unto man to be bought at the price of labour, to be attained by a few easy and slothful observances. Deficiencies in these knowledges I will report none, other than the general deficience, that it is not known how much of them is verity, and how much vanity.[140]

XII 1.

The Knowledge which respecteth the faculties of the mind of man is of two kinds; the one respecting his Understanding and Reason, and the other his Will, Appetite, and Affection; whereof the former produceth Position or Decree, the latter Action or Execution. It is true that the Imagination is an agent or *nuncius,* in both provinces, both the judicial and the ministerial. For Sense sendeth over to Imagination before Reason have judged: and Reason sendeth over to Imagination before the decree can be acted: for Imagination ever precedeth Voluntary Motion. Saving that this Janus of Imagination hath differing faces: for the face towards Reason hath the print of Truth, but the face towards Action hath the print of Good; which nevertheless are faces,

Quales decet esse sororum.[141]

Neither is the Imagination simply and only a messenger; but is invested with, or at least wise usurpeth no small authority in itself, besides the duty of the message. For it was well said by Aristotle, *That the mind hath over the body that commandment, which the lord hath over a bondman; but that reason hath over the imagination that commandment which a magistrate hath over a free citizen;*[142] who may come also to rule in his turn. For we see that, in matters of Faith and Religion, we raise our Imagination above our Reason; which is the cause why Religion sought ever access to the mind by similitude, types, parables, visions, dreams. And again, in all persuasions that are wrought by eloquence, and other impressions of like nature, which do paint and disguise the true appearance of things, the chief recommendation unto Reason is from the Imagination.[143] Nevertheless, because I find not any science that doth properly or fitly pertain to the Imagination, I see no cause to alter the former division. For as for poesy, it is rather a pleasure or play of Imagination, than a work or duty thereof. And if it be a work, we speak not now of such parts of learning as the Imagination produceth, but of such sciences as handle and consider of the Imagination; no more than we shall speak now of such knowledges as reason produceth, for that extendeth to all philosophy, but of such knowledges as do handle and inquire the faculty of reason: so as poesy had its true place. As for the power of the Imagination in nature, and the manner of fortifying the same, we have mentioned it in the doctrine *De Anima,* whereunto it most fitly belongeth. And lastly, for Imaginative or Insinuative Reason, which is the subject of Rhetoric, we think it best to refer it to the Arts of Reason. So therefore we content ourselves with the former division, that human philosophy, which respecteth the faculties of the mind of man, hath two parts, rational and moral.

2.   The part of human philosophy which is rational, is of all knowledges, to the most wits, the least delightful; and seemeth but a net of subtlety and spinosity. For as it was truly said, that knowledge is *Pabulum animi,*[144] so in the nature of men's appetite to this food, most men are of the taste and stomach of the Israelites in the desert, that would fain have returned *ad ollas carnium,*[145] and were weary of manna; which, though it were celestial, yet seemed less nutritive and comfortable. So generally men taste well knowledges that are drenched in flesh and blood, civil history, morality, policy, about the which men's affections, praises, fortunes do turn and are conversant; but this same *lumen siccum* doth parch and offend most men's watery and soft natures. But to speak truly of things as they are in worth, Rational Knowledges are the keys of all other arts, for as Aristotle saith, aptly and elegantly, *That the hand is the instrument of instruments, and the mind is the form of forms:*[146] so these be truly said to be the art of arts: neither do they only direct, but likewise confirm and strengthen: even as the habit of shooting doth not only enable to shoot a nearer shoot, but also to draw a stronger bow.

3.   The *Arts intellectual* are four in number; divided according to the ends whereunto they are referred: for man's labour is to *invent* that which is sought or propounded; or to *judge* that which is invented; or to *retain* that which is judged; or to *deliver over* that which is retained. So as the

arts must be four: *Art* of *Inquiry* or *Invention: Art* of *Examination* or *Judgment: Art* of *Custody* or *Memory:* and *Art* of *Elocution* or *Tradition.*

XIII

1. *Invention* is of two kinds, much differing: the one of *Arts* and *Sciences;* and the other of *Speech* and *Arguments.* The former of these I do report deficient; which seemeth to me to be such a deficience as if in the making of an inventory touching the estate of a defunct it should be set down *that there is no ready money.* For as money will fetch all other commodities, so this knowledge is that which should purchase all the rest. And like as the West Indies had never been discovered if the use of the mariner's needle had not been first discovered, though the one be vast regions, and the other a small motion; so it cannot be found strange if sciences be no further discovered, if the art itself of invention and discovery hath been passed over.

2. That this part of knowledge is wanting, to my judgment standeth plainly confessed; for first, Logic doth not pretend to invent sciences, or the axioms of sciences, but passeth it over with a *Cuique in sua arte credendum.*[147] And Celsus acknowledgeth it gravely, speaking of the Empirical and dogmatical sects of physicians, *That medicines and cures were first found out, and then after the reasons and causes were discoursed; and not the causes first found out, and by light from them the medicines and cures discovered.*[148] And Plato, in his *Theaetetus,* noteth well, *That particulars are infinite, and the higher generalities give no sufficient direction: and that the pith of all sciences, which maketh the artsman differ from the inexpert, is in the middle propositions, which in every particular knowledge are taken from tradition and experience.*[149] And therefore we see, that they which discourse of the inventions and originals of things, refer them rather to chance than to art, and rather to beasts, birds, fishes, serpents, than to men.

> Dictamnum genetrix Cretaea carpit ab Ida,
> Puberibus caulem foliis et flore comantem
> Purpureo; non illa feris incognita capris
> Gramina, cum tergo volucres haesere sagittae.[150]

So that it was no marvel, the manner of antiquity being to consecrate inventors, that the Egyptians had so few human idols in their temples, but almost all brute:

> Omnigenumque Deum monstra, et latrator Anubis,
> Contra Neptunum, et Venerem, contraque Minervam, etc.[151]

And if you like better the tradition of the Grecians, and ascribe the first inventions to men; yet you will rather believe that Prometheus first struck the flints, and marvelled at the spark, than that when he first struck the flints he expected the spark: and therefore we see the West Indian Prometheus[152] had no intelligence with the European, because of the rareness with them of flint, that gave the first occasion. So as it should seem, that hitherto men are rather beholden to a wild goat for surgery, or to a nightingale for music, or to the ibis for some part of physic, or to the pot-lid that flew open for artillery, or generally to chance,[153] or anything else, than to logic, for the invention of arts and sciences. Neither is the form of invention which Virgil describeth much other:

> Ut varias usus meditando extunderet artes
> Paulatim.[154]

For if you observe the words well, it is no other method than that which brute beasts are capable of, and do put in ure; which is a perpetual intending or practising some one thing, urged and imposed by an absolute necessity of conservation of being; for so Cicero saith very truly, *Usus uni rei deditus et naturam et artem saepe vincit.*[155] And therefore if it be said of men,

> Labor omnia vincit
> Improbus, et duris urgens in rebus egestas![156]

it is likewise said of beasts, *Quis psittaco docuit suum χαῖρε?*[157] Who taught the raven in a drought to throw pebbles into a hollow tree, where she spied water, that the water might rise so as she might come to it? Who taught the bee to sail through such a vast sea of air, and to find the way from a field in flower a great way off to her hive? Who taught the ant to bite every grain of corn that she burieth in her hill, lest it should take root and grow? Add then the word *extundere,* which importeth the extreme difficulty, and the word *paulatim,* which importeth the extreme slowness, and we are where we were, even amongst the Egyptians' gods; there being little left to the faculty of reason, and nothing to the duty of art, for matter of invention.

3.   Secondly, the Induction which the Logicians speak of, and which seemeth familiar with Plato, whereby the Principles of Sciences may be pretended to be invented, and so the middle propositions by derivation from the Principles; their form of induction, I say, is utterly vicious and incompetent: wherein their error is the fouler, because it is the duty of Art to perfect and exalt Nature; but they contrariwise have wronged, abused, and traduced Nature. For he that shall attentively observe how the mind doth gather this excellent dew of knowledge, like unto that which the poet speaketh of, *Aerei mellis caelestia dona,*[158] distilling and contriving it out of particulars natural and artificial, as the flowers of the field and garden, shall find that the mind of herself by nature doth manage and act an induction much better than they describe it. For to conclude upon an enumeration of particulars, without instance contradictory, is no conclusion, but a conjecture; for who can assure, in many subjects, upon those particulars which appear of a side, that there are not other on the contrary side which appear not? As if Samuel should have rested upon those sons of Jesse[159] which were brought before him, and failed of David, which was in the field.[160] And this form, to say truth, is so gross, as it had not been possible for wits so subtle as have managed these things to have offered it to the world, but that they hasted to their theories and dogmaticals, and were imperious and scornful towards particulars; which their manner was to use but as *lictores* and *viatores,* for sergeants and whifflers, *ad summovendam turbam,* to make way and make room for their opinions, rather than in their true use and service. Certainly it is a thing may touch a man with a religious wonder, to see how the footsteps of seducement are the very same in divine and human truth: for as in divine truth man cannot endure to become as a child; so in human, they reputed the attending the inductions whereof we speak, as if it were a second infancy or childhood.

4.   Thirdly, allow some principles or axioms were rightly induced, yet nevertheless certain it is that middle propositions cannot be de-

duced from them in subject of nature[161] by syllogism, that is, by touch and reduction of them to principles in a middle term. It is true that in sciences popular, as moralities, laws, and the like, yea, and divinity (because it pleaseth God to apply himself to the capacity of the simplest), that form may have use; and in natural philosophy likewise, by way of argument or satisfactory reason, *Quae assensum parit, operis effoeta est:*[162] but the subtlety of nature and operations will not be enchained in those bonds: for arguments consist of propositions, and propositions of words; and words are but the current tokens or marks[163] of popular notions of things; which notions, if they be grossly and variably collected out of particulars, it is not the laborious examination either of consequence of arguments, or of the truth of propositions, that can ever correct that error, being, as the physicians speak, in the first digestion: and therefore it was not without cause, that so many excellent philosophers became Sceptics and Academics, and denied any certainty of knowledge or comprehension; and held opinion that the knowledge of man extended only to appearances and probabilities. It is true that in Socrates it was supposed to be but a form of irony, *Scientiam dissimulando simulavit,*[164] for he used to disable his knowledge, to the end to enhance his knowledge: like the humour of Tiberius in his beginnings, that would reign, but would not acknowledge so much:[165] and in the later Academy, which Cicero embraced, this opinion also of *acatalepsia,*[166] I doubt, was not held sincerely: for that all those which excelled in copie of speech seem to have chosen that sect, as that which was fittest to give glory to their eloquence and variable discourses; being rather like progresses of pleasure, than journeys to an end. But assuredly many scattered in both Academies did hold it in subtlety and integrity: but here was their chief error; they charged the deceit upon the senses; which in my judgment, notwithstanding all their cavilations, are very sufficient to certify and report truth though not always immediately, yet by comparison, by help of instrument, and by producing and urging such things as are too subtle for the sense to some effect comprehensible by the sense, and other like assistance. But they ought to have charged the deceit upon the weakness of the intellectual powers, and upon the manner of collecting and concluding upon the reports of the senses. This I speak, not to disable the mind of man, but to stir it up to seek help: for no man, be he never so cunning or practised, can make a straight line or perfect circle by steadiness of hand, which may be easily done by help of a ruler or compass.

5.   This part of invention, concerning the invention of sciences, I purpose, if God give me leave, hereafter to propound, having digested it into two parts; whereof the one I term *experientia literata,* and the other *interpretatio naturae:* the former being but a degree and rudiment of the latter. But I will not dwell too long, nor speak too great upon a promise.[167]

6.   The invention of speech or argument is not properly an invention, for to *invent* is to discover that we know not, and not to recover or resummon that which we already know: and the use of this invention is no other but out of the knowledge whereof our mind is already possessed to draw forth or call before us that which may be pertinent to the purpose which we take into our consideration. So as to speak truly, it is no invention, but a remembrance or suggestion, with an application; which is the cause why the schools do place it after judgment, as subsequent and not precedent. Nevertheless, because we do account it a chase as well of deer in an inclosed park as in a forest at large, and that it hath already obtained the name, let it be called invention: so as it be perceived and discerned, that the scope and end of this invention is readiness and present use of our knowledge, and not addition or amplification thereof.

7.   To procure this ready use of knowledge there are two courses, Preparation and Suggestion. The former of these seemeth scarcely a part of knowledge, consisting rather of diligence than of any artificial erudition. And herein Aristotle wittily, but hurtfully, doth deride the Sophists near his time, saying, *They did as if one that professed the art of shoe-making should not teach how to make a shoe, but only exhibit in a readiness a number of shoes of all fashions and sizes.*[168] But yet a man might reply, that if a shoemaker should have no shoes in his shop, but only work as he is bespoken, he should be weakly customed. But our Saviour, speaking of divine knowledge, saith, *that the kingdom of heaven is like a good householder, that bringeth forth both new and old store:*[169] and we see the ancient writers of Rhetoric do give it in precept, that pleaders should have the Places, whereof they have most continual use,

ready handled in all the variety that may be; as that, to speak for the literal interpretation of the law against equity, and contrary; and to speak for presumptions and inferences against testimony, and contrary. And Cicero himself, being broken unto it by great experience, delivereth it plainly, that whatsoever a man shall have occasion to speak of, if he will take the pains, he may have it in effect premeditate, and handled, *in thesi;*[170] so that when he cometh to a particular he shall have nothing to do, but to put to names and times and places, and such other circumstances of individuals. We see likewise the exact diligence of Demosthenes; who, in regard of the great force that the entrance and access into causes hath to make a good impression, had ready framed a number of prefaces for orations and speeches. All which authorities and precedents may overweigh Aristotle's opinion, that would have us change a rich wardrobe for a pair of shears.

8.　　But the nature of the collection of this provision or preparatory store, though it be common both to Logic and Rhetoric, yet having made an entry of it here, where it came first to be spoken of, I think fit to refer over the further handling of it to Rhetoric.

9.　　The other part of invention, which I term suggestion, doth assign and direct us to certain *marks,* or *places,* which may excite our mind to return and produce such knowledge as it hath formerly collected, to the end we may make use thereof. Neither is this use, truly taken, only to furnish argument to dispute probably with others, but likewise to minister unto our judgment to conclude aright within ourselves. Neither may these Places serve only to apprompt our invention, but also to direct our inquiry. For a faculty of wise interrogating is half a knowledge. For as Plato saith, *Whosoever seeketh, knoweth that which he seeketh for in a general notion: else how shall he know it when he hath found it?*[171] and therefore the larger your anticipation is, the more direct and compendious is your search. But the same Places which will help us what to produce of that which we know already, will also help us, if a man of experience were before us, what questions to ask; or, if we have books and authors to instruct us, what points to search and revolve; so as I cannot report that this part of invention, which is that which the schools call *Topics,* is deficient.[172]

10.　　Nevertheless, *Topics* are of two sorts, general and special.[173] The general we have spoken to; but the particular hath been touched by some, but rejected generally as inartificial and variable. But leaving the humour which hath reigned too much in the schools, which is, to be vainly subtle in a few things which are within their command, and to reject the rest; I do receive particular Topics (that is, places or directions of invention and inquiry in every particular knowledge) as things of great use, being mixtures of Logic with the matter of sciences; for in these it holdeth, *ars inveniendi adolescit cum inventis;*[174] for as in going of a way, we do not only gain that part of the way which is passed, but we gain the better sight of that part of the way which remaineth: so every degree of proceeding in a science giveth a light to that which followeth; which light if we strengthen by drawing it forth into questions or places of inquiry, we do greatly advance our pursuit.[175]

XIV 1. Now we pass unto the arts of Judgment, which handle

the natures of Proofs and Demonstrations; which as to Induction hath a coincidence with Invention. For in all inductions, whether in good or vicious form, the same action of the mind which inventeth, judgeth; all one as in the sense. But otherwise it is in proof by syllogism; for the proof being not immediate, but by mean, the invention of the mean is one thing, and the judgment of the consequence is another; the one exciting only, the other examining. Therefore for the real and exact form of judgment, we refer ourselves to that which we have spoken of *interpretation of nature.*[176]

2.   For the other judgment by Syllogism, as it is a thing most agreeable to the mind of man, so it hath been vehemently and excellently laboured; for the nature of man doth extremely covet to have somewhat in his understanding fixed and unmoveable, and as a rest and support of the mind. And therefore as Aristotle endeavoureth to prove, that in all motion there is some point quiescent;[177] and as he elegantly expoundeth the ancient fable of Atlas, that stood fixed, and bare up the heaven from falling, to be meant of the poles or axle-tree of heaven, whereupon the conversion is accomplished; so assuredly men have a desire to have an *Atlas* or axle-tree within to keep them from fluctuation, which is like to a perpetual peril of falling; therefore men did hasten

to set down some principles about which the variety of their disputations might turn.

3.   So then this art of Judgment is but the reduction of propositions to principles in a middle term: the principles to be agreed by all and exempted from argument; the middle term to be elected at the liberty of every man's invention; the reduction to be of two kinds, direct and inverted; the one when the proposition is reduced to the principle, which they term a *probation ostensive:* the other, when the contradictory of the proposition is reduced to the contradictory of the principle, which is that which they call *per incommodum,* or *pressing an absurdity;* the number of middle terms to be as the proposition standeth degrees more or less removed from the principle.[178]

4.   But this art hath two several methods of doctrine, the one by way of direction, the other by way of caution: the former frameth and setteth down a true form of consequence, by the variations and deflections from which errors and inconsequences may be exactly judged. Toward the composition and structure of which form, it is incident to handle the parts thereof, which are propositions, and the parts of propositions, which are simple words: and this is that part of Logic which is comprehended in the *Analytics.*

5.   The second method of doctrine was introduced for expedite use and assurance sake; discovering the more subtle forms of sophisms and illaqueations with their redargutions, which is that which is termed *elenches.* For although in the more gross sorts of fallacies it happeneth, as Seneca maketh the comparison well, as in juggling feats, which, though we know not how they are done, yet we know well it is not as it seemeth to be;[179] yet the more subtle sort of them doth not only put a man besides his answer, but doth many times abuse his judgment.

6.   This part concerning *elenches* is excellently handled by Aristotle in precept, but more excellently by Plato in example: not only in the persons of the Sophists, but even in Socrates himself, who, professing to affirm nothing, but to infirm that which was affirmed by another, hath exactly expressed all the forms of objection, fallacy, and redargution.[180] And although we have said that the use of this doctrine is for redargution, yet it is manifest the degenerate and corrupt use is for caption and contradiction, which passeth for a great faculty, and no

doubt is of very great advantage: though the difference be good which was made between orators and sophisters, that the one is as the greyhound which hath his advantage in the race, and the other as the hare which hath her advantage in the turn, so as it is the advantage of the weaker creature.

7.   But yet further, this doctrine of *elenches* hath a more ample latitude and extent than is perceived; namely, unto divers parts of knowledge; whereof some are laboured and others omitted. For first, I conceive, though it may seem at first somewhat strange, that that part which is variably referred, sometimes to logic, sometimes to metaphysics, touching the common adjuncts of essences, is but an *elench;* for the great sophism of all sophisms being equivocation, or ambiguity of words and phrase (especially of such words as are most general, and intervene in every inquiry), it seemeth to me that the true and fruitful use, leaving vain subtleties and speculations, of the inquiry of *majority, minority, priority, posteriority, identity, diversity, possibility, act, totality, parts, existence, privation,* and the like, are but wise cautions against the ambiguities of speech. So again the distribution of things into certain tribes, which we call *categories* or *predicaments,* are but cautions against the confusion of *definitions* and *divisions.*[181]

8.   Secondly, there is a seducement that worketh by the strength of the impression, and not by the subtlety of the illaqueation; not so much perplexing the reason, as overruling it by power of the imagination. But this part I think more proper to handle when I shall speak of rhetoric.

9.   But lastly, there is yet a much more important and profound kind of fallacies in the mind of man, which I find not observed or inquired at all,[182] and think good to place here, as that which of all others appertaineth most to rectify judgment: the force whereof is such, as it doth not dazzle or snare the understanding in some particulars, but doth more generally and inwardly infect and corrupt the state thereof. For the mind of man is far from the nature of a clear and equal glass, wherein the beams of things should reflect according to their true incidence; nay, it is rather like an enchanted glass, full of superstition and imposture, if it be not delivered and reduced. For this purpose, let us consider the false appearances that are imposed upon us by the gen-

eral nature of the mind,[183] beholding them in an example or two; as first, in that instance which is the root of a superstition, namely, *That to the nature of the mind of all men it is consonant for the affirmative or active to affect more than the negative or privative:* so that a few times hitting or presence, countervails oft-times failing or absence; as was well answered by Diagoras to him that showed him in Neptune's temple the great number of pictures of such as had escaped shipwreck, and had paid their vows to Neptune, saying, *Advise now, you that think it folly to invoke Neptune in tempest: Yea, but,* saith Diagoras, *where are they painted that are drowned?*[184] Let us behold it in another instance, namely, *That the spirit of man, being of an equal and uniform substance, doth usually suppose and feign in nature a greater equality and uniformity than is in truth.* Hence it cometh, that the mathematicians cannot satisfy themselves except they reduce the motions of the celestial bodies to perfect circles, rejecting spiral lines, and labouring to be discharged of eccentrics.[185] Hence it cometh, that whereas there are many things in nature, as it were *monodica, sui juris;*[186] yet the cogitations of man do feign unto them *relatives, parallels,* and *conjugates,* whereas no such thing is; as they have feigned an element of fire, to keep square with earth, water, and air, and the like: nay, it is not credible, till it be opened, what a number of fictions and fancies the similitude of human actions and arts, together with the making of man *communis mensura,* have brought into natural philosophy; not much better than the heresy of the Anthropomorphites,[187] bred in the cells of gross and solitary monks, and the opinion of Epicurus, answerable to the same in heathenism, who supposed the Gods to be of human shape. And therefore Velleius the Epicurean needed not to have asked why God should have adorned the heavens with stars, as if he had been an *aedilis,* one that should have set forth some magnificent shows or plays.[188] For if that great Workmaster had been of a human disposition, he would have cast the stars into some pleasant and beautiful works and orders, like the frets in the roofs of houses; whereas one can scarce find a posture in square, or triangle, or straight line, amongst such an infinite number; so differing a harmony there is between the spirit of man and the spirit of nature.

10.   Let us consider again the false appearances imposed upon us by every man's own individual nature and custom,[189] in that feigned supposition that Plato[190] maketh of the cave: for certainly if a child were continued in a grot or cave under the earth until maturity of age, and came suddenly abroad, he would have strange and absurd imaginations. So in like manner, although our persons live in the view of heaven, yet our spirits are included in the caves of our own complexions and customs, which minister unto us infinite errors and vain opinions, if they be not recalled to examination. But hereof we have given many examples in one of the errors, or peccant humours, which we ran briefly over in our first book.

11.   And lastly, let us consider the false appearances that are imposed upon us by words, which are framed and applied according to the conceit and capacities of the vulgar sort: and although we think we govern our words, and prescribe it well, *loquendum ut vulgus, sentiendum ut sapientes;* yet certain it is that words, as a Tartar's bow, do shoot back upon the understanding of the wisest, and mightily entangle and pervert the judgment. So as it is almost necessary in all controversies and disputations to imitate the wisdom of the mathematicians, in setting down in the very beginning the definitions of our words and terms that others may know how we accept and understand them, and whether they concur with us or no. For it cometh to pass for want of this that we are sure to end there where we ought to have begun, which is, in questions and differences about words. To conclude therefore, it must be confessed that it is not possible to divorce ourselves from these fallacies and false appearances, because they are inseparable from our nature and condition of life; so yet nevertheless the caution of them (for all *elenches,* as was said, are but cautions) doth extremely import the true conduct of human judgment. The particular *elenches* or cautions against these three false appearances, I find altogether deficient.

12.   There remaineth one part of judgment of great excellency, which to mine understanding is so slightly touched, as I may report that also deficient; which is the application of the differing kinds of proofs to the differing kinds of subjects; for there being but four kinds of demonstrations, that is, *by the immediate consent of the mind or sense,* by *induction,* by *syllogism,* and by *congruity* (which is that which Aristotle calleth *demonstration in orb or circle,*[191] and not *a notioribus*); every of

these hath certain subjects in the matter of sciences, in which respectively they have chiefest use; and certain others, from which respectively they ought to be excluded; and the rigour and curiosity in requiring the more severe proofs in some things, and chiefly the facility in contenting ourselves with the more remiss proofs in others, hath been amongst the greatest causes of detriment and hinderance to knowledge. The distributions and assignations of demonstrations, according to the analogy of sciences, I note as deficient.

XV 1.   The custody or retaining of knowledge is either in writing or memory; whereof writing hath two parts, the nature of the character, and the order of the entry; for the art of characters, or other visible notes of words or things, it hath nearest conjugation with grammar; and therefore I refer it to the due place: for the disposition and collocation of that knowledge which we preserve in writing, it consisteth in a good digest of common-places; wherein I am not ignorant of the prejudice imputed to the use of common-place books, as causing a retardation of reading, and some sloth or relaxation of memory. But because it is but a counterfeit thing in knowledges to be forward and pregnant, except a man be deep and full, I hold the entry of common-places to be a matter of great use and essence in studying, as that which assureth copie of invention, and contracteth judgment to a strength. But this is true, that of the methods of common-places that I have seen, there is none of any sufficient worth; all of them carrying merely the face of a school, and not of a world; and referring to vulgar matters and pedantical divisions, without all life or respect to action.

2.   For the other principal part of the custody of knowledge, which is Memory, I find that faculty in my judgment weakly inquired of. An art[192] there is extant of it; but it seemeth to me that there are better precepts than that art, and better practices of that art than those received. It is certain the art, as it is, may be raised to points of ostentation prodigious: but in use, as it is now managed, it is barren (not burdensome, nor dangerous to natural memory, as is imagined, but barren), that is, not dexterous to be applied to the serious use of business and occasions. And therefore I make no more estimation of repeating a great

number of names or words upon once hearing, or the pouring forth of a number of verses or rhymes, *ex tempore,* or the making of a satirical simile of everything, or the turning of everything to a jest, or the falsifying or contradicting of everything by cavil, or the like (whereof in the faculties of the mind there is great copie, and such as by device and practice may be exalted to an extreme degree of wonder), than I do of the tricks of tumblers, *funambuloes, baladines:*[193] the one being the same in the mind that the other is in the body, matters of strangeness without worthiness.

3.   This art of memory is but built upon two intentions; the one prenotion, the other emblem. Prenotion dischargeth the indefinite seeking of that we would remember, and directeth us to seek in a narrow compass, that is, somewhat that hath congruity with our place of memory. Emblem reduceth conceits intellectual to images sensible, which strike the memory more; out of which axioms may be drawn much better practice than that in use; and besides which axioms, there are divers more touching help of memory, not inferior to them. But I did in the beginning distinguish, not to report those things deficient, which are but only ill managed.

XVI   1.   There remaineth the fourth kind of rational knowledge, which is transitive, concerning the expressing or transferring our knowledge to others; which I will term by the general name of tradition or delivery. Tradition hath three parts; the first concerning the organ of tradition: the second concerning the method of tradition; and the third concerning the illustration of tradition.

2.   For the organ of tradition, it is either speech or writing: for Aristotle saith well, *Words are the images of cogitations, and letters are the images of words;*[194] but yet it is not of necessity that cogitations be expressed by the medium of words. For *whatsoever is capable of sufficient differences, and those perceptible by the sense, is in nature competent to express cogitations.* And therefore we see in the commerce of barbarous people, that understand not one another's language, and in the practice of divers that are dumb and deaf, that men's minds are expressed in gestures, though not exactly, yet to serve the turn. And

we understand further, that it is the use of China, and the kingdoms of the high Levant,[195] to write in characters real, which express neither letters nor words in gross, but things or notions; insomuch as countries and provinces, which understand not one another's language, can nevertheless read one another's writings, because the characters are accepted more generally than the languages do extend; and therefore they have a vast multitude of characters, as many, I suppose, as radical words.

3.　These notes of cogitations are of two sorts; the one when the note hath some similitude or congruity with the notion: the other *ad placitum,* having force only by contract or acceptation. Of the former sort are hieroglyphics and gestures. For as to hieroglyphics, things of ancient use, and embraced chiefly by the Egyptians, one of the most ancient nations, they are but as continued impresses and emblems. And as for gestures, they are as transitory hieroglyphics, and are to hieroglyphics as words spoken are written, in that they abide not; but they have evermore, as well as the other, an affinity with the things signified: as Periander, being consulted with how to preserve a tyranny newly usurped, bid the messenger attend and report what he saw him do; and went into his garden and topped all the highest flowers: signifying, that it consisted in the cutting off and keeping low of the nobility and grandees.[196] *Ad placitum,* are the characters real before mentioned, and words: although some have been willing by curious inquiry, or rather by apt feigning to have derived imposition of names from reason and intendment; a speculation elegant, and, by reason it searcheth into antiquity, reverent; but sparingly mixed with truth, and of small fruit. This portion of knowledge, touching the notes of things and cogitations in general, I find not inquired but deficient. And although it may seem of no great use, considering that words and writings by letter do far excel all the other ways; yet because this part concerneth, as it were, the mint of knowledge (for words are the tokens current and accepted for conceits, as moneys are for values, and that it is fit men be not ignorant that moneys may be of another kind than gold and silver), I thought good to propound it to better inquiry.

4.　Concerning speech and words, the consideration of them hath produced the science of grammar: for man still striveth to reintegrate himself in those benedictions, from which by his fault he hath been deprived; and as he hath striven against the first general curse by the invention of all other arts, so hath he sought to come forth of the second general curse, which was the confusion of tongues, by the art of grammar; whereof the use in a mother tongue is small, in a foreign tongue more; but most in such foreign tongues as have ceased to be vulgar tongues, and are turned only to learned tongues. The duty of it is of two natures; the one popular, which is for the speedy and perfect attaining languages, as well for intercourse of speech as for understanding of authors; the other philosophical, examining the power and nature of words, as they are the footsteps and prints of reason: which kind of analogy between words and reason is handled *sparsim,* brokenly, though not entirely; and therefore I cannot report it deficient, though I think it very worthy to be reduced into a science by itself.

5.   Unto grammar also belongeth, as an appendix, the consideration of the accidents of words; which are measure, sound, and elevation or accent, and the sweetness and harshness of them; whence hath issued some curious observations in rhetoric, but chiefly poesy, as we consider it, in respect of the verse and not of the argument; wherein though men in learned tongues do tie themselves to the ancient measures, yet in modern languages it seemeth to me as free to make new measures of verses as of dances: for a dance is a measured pace, as a verse is a measured speech. In these things the sense is better judge than the art;

> Coenae fercula nostrae
> Mallem convivis quam placuisse cocis.[197]

And of the servile expressing antiquity in an unlike and an unfit subject, it is well said, *Quod tempore antiquum videtur, id incongruitate est maxime novum.*[198]

6.   For ciphers, they are commonly in letters or alphabets but may be in words. The kinds of ciphers, besides the simple ciphers, with changes, and intermixtures of nulls and non-significants, are many, according to the nature or rule of the infolding, wheel-ciphers, key-ciphers, doubles, etc.[199] But the virtues of them, whereby they are to be preferred, are three; that they be not laborious to write and read; that they be impossible to decipher; and, in some cases, that they be with-

out suspicion. The highest degree whereof is to write *omnia per omnia;* which is undoubtedly possible, with a proportion quintuple at most of the writing infolding to the writing infolded, and no other restraint whatsoever. This art of ciphering hath for relative an art of deciphering, by supposition unprofitable, but, as things are, of great use. For suppose that ciphers were well managed, there be multitudes of them which exclude the decipherer. But in regard of the rawness and unskilfulness of the hands through which they pass, the greatest matters are many times carried in the weakest ciphers.

7.   In the enumeration of these private and retired arts, it may be thought I seek to make a great muster-roll of sciences, naming them for show and ostentation, and to little other purpose. But let those which are skilful in them judge whether I bring them in only for appearance, or whether in that which I speak of them, though in few words, there be not some seed of proficience. And this must be remembered, that as there be many of great account in their countries and provinces, which, when they come up to the seat of the estate, are but of mean rank and scarcely regarded; so these arts, being here placed with the principal and supreme sciences, seem petty things; yet to such as have chosen them to spend their labours and studies in them, they seem great matters.

XVII   1.   For the Method of Tradition, I see it hath moved a controversy in our time.[200] But as in civil business, if there be a meeting, and men fall at words, there is commonly an end of the matter for that time, and no proceeding at all; so in learning, where there is much controversy, there is many times little inquiry. For this part of knowledge of Method seemeth to me so weakly inquired as I shall report it deficient.

2.   Method hath been placed, and that not amiss, in Logic, as a part of Judgment;[201] for as the doctrine of Syllogisms comprehendeth the rules of Judgment upon that which is invented, so the doctrine of Method containeth the rules of Judgment upon that which is to be delivered; for Judgment precedeth Delivery, as it followeth Invention. Neither is the Method or the nature of the tradition material only to the use of knowledge, but likewise to the progression of knowledge:

for since the labour and life of one man cannot attain to perfection of knowledge, the wisdom of the tradition is that which inspireth the felicity of continuance and proceeding. And therefore the most real diversity of method, is of Method referred to use, and Method referred to progression: whereof the one may be termed Magistral, and the other of Probation.

3.   The latter whereof seemeth to be *via deserta et interclusa*. For as knowledges are now delivered, there is a kind of contract of error between the deliverer and the receiver: for he that delivereth knowledge, desireth to deliver it in such form as may be best believed, and not as may be best examined; and he that receiveth knowledge, desireth rather present satisfaction, than expectant inquiry; and so rather not to doubt, than not to err: glory making the author not to lay open his weakness, and sloth making the disciple not to know his strength.

4.   But knowledge that is delivered as a thread to be spun on, ought to be delivered and intimated, if it were possible, in the same method wherein it was invented: and so is it possible of knowledge induced. But in this same anticipated and prevented knowledge, no man knoweth how he came to the knowledge which he hath obtained. But yet nevertheless, *secundum majus et minus,* a man may revisit and descend unto the foundations of his knowledge and consent; and so transplant it into another, as it grew in his own mind. For it is in knowledges as it is in plants: if you mean to use the plant, it is no matter for the roots; but if you mean to remove it to grow, then it is more assured to rest upon roots than slips: so the delivery of knowledges, as it is now used, is as of fair bodies of trees without the roots; good for the carpenter, but not for the planter. But if you will have sciences grow, it is less matter for the shaft or body of the tree, so you look well to the taking up of the roots: of which kind of delivery the method of the mathematics, in that subject, hath some shadow: but generally I see it neither put in use[202] nor put in inquisition: and therefore note it for deficient.

5.   Another diversity of Method there is, which hath some affinity with the former, used in some cases by the discretion of the ancients, but disgraced since by the impostures of many vain persons, who have made it as a false light for their counterfeit merchandises; and that is, enigmatical and disclosed.[203] The pretence whereof is, to remove the

vulgar capacities from being admitted to the secrets of knowledges, and to reserve them to selected auditors, or wits of such sharpness as can pierce the veil.

6. Another diversity of Method, whereof the consequence is great, is the delivery of knowledge in Aphorisms, or in Methods; wherein we may observe that it hath been too much taken into custom, out of a few axioms or observations upon any subject, to make a solemn and formal art, filling it with some discourses, and illustrating it with examples, and digesting it into a sensible Method. But the writing in Aphorisms hath many excellent virtues, where to the writing in Method doth not approach.

7. For first, it trieth the writer, whether he be superficial or solid: for Aphorisms, except they should be ridiculous, cannot be made but of the pith and heart of sciences; for discourse of illustration is cut off: recitals of examples are cut off; discourse of connection and order is cut off; descriptions of practice are cut off. So there remaineth nothing to fill the Aphorisms but some good quantity of observation: and therefore no man can suffice, nor in reason will attempt to write Aphorisms, but he that is sound and grounded. But in Methods,

> Tantum series juncturaque pollet,
> Tantum de medio sumptis accedit honoris:[204]

as a man shall make a great show of an art, which, if it were disjointed, would come to little. Secondly, methods are more fit to win consent or belief, but less fit to point to action; for they carry a kind of demonstration in orb or circle, one part illuminating another, and therefore satisfy; but particulars, being dispersed, do best agree with dispersed directions. And lastly, Aphorisms, representing a knowledge broken, do invite men to inquire further; whereas Methods, carrying the show of a total, do secure men, as if they were at furthest.

8. Another diversity of Method, which is likewise of great weight, is the handling of knowledge by assertions and their proofs, or by questions and their determinations; the latter kind whereof, if it be immoderately followed, is as prejudicial to the proceeding of learning, as it is to the proceeding of an army to go about to besiege every little fort or hold. For if the field be kept, and the sum of the enterprise

pursued, those smaller things will come in of themselves: indeed a man would not leave some important piece enemy at his back.[205] In like manner, the use of confutation in the delivery of sciences ought to be very sparing; and to serve to remove strong preoccupations and pre-judgments, and not to minister and excite disputations and doubts.

9.   Another diversity of Method is, according to the subject or matter which is handled; for there is a great difference in delivery of the mathe-matics, which are most abstracted of knowledges, and policy, which is the most immersed: and howsoever contention hath been moved, touch-ing a uniformity of method in multiformity of matter, yet we see how that opinion, besides the weakness of it, hath been of ill desert towards learning, as that which taketh the way to reduce learning to certain empty and barren generalities; being but the very husks and shells of sciences, all the kernel being forced out and expulsed with the torture and press of the Method. And therefore as I did allow well of particular topics for invention, so I do allow likewise of particular Methods of tradition.

10.   Another diversity of judgment[206] in the delivery and teaching of knowledge is according unto the light and presuppositions of that which is delivered; for that knowledge which is new, and foreign from opinions received, is to be delivered in another form than that that is agreeable[207] and familiar; and therefore Aristotle, when he thinks to tax Democritus, doth in truth commend him, where he saith, *If we shall indeed dispute, and not follow after similitudes, etc.*[208] For those whose conceits are seated in popular opinions, need only but to prove or dispute; but those whose conceits are beyond popular opinions, have a double labour; the one to make themselves conceived, and the other to prove and demonstrate: so that it is of necessity with them to have recourse to similitudes and translations to express themselves. And therefore in the infancy of learning, and in rude times, when those conceits which are now trivial were then new, the world was full of parables and similitudes; for else would men either have passed over without mark, or else rejected for paradoxes, that which was offered, before they had understood or judged. So in divine learning, we see how frequent parables and tropes are: for it is a rule, that what-soever science is not consonant to presuppositions, must pray in aid of similitudes.

11. There be also other diversities of Methods vulgar and received: as that of Resolution or Analysis, of Constitution or Systasis, of Concealment of Cryptic, etc., which I do allow well of, though I have stood upon those which are least handled and observed. All which I have remembered to this purpose, because I would erect and constitute one general inquiry, which seems to me deficient, touching the Wisdom of Tradition.

12. But unto this part of knowledge, concerning Method, doth further belong not only the architecture of the whole frame of a work, but also the several beams and columns thereof; not as to their stuff, but as to their quantity and figure. And therefore Method considereth not only the disposition of the argument or subject, but likewise the propositions: not as to their truth or matter, but as to their limitation and manner. For herein Ramus merited better a great deal in reviving the good rules of propositions, *Καθόλου πρῶτον κατὰ παντός*, etc., than he did in introducing the canker of epitomes;[209] and yet (as it is the condition of human beings that, according to the ancient fables, *the most precious things have the most pernicious keepers*) it was so, that the attempt of the one made him fall upon the other. For he had need be well conducted that should design to make axioms convertible, if he make them not withal circular, and non-promovent, or incurring into themselves; but yet the intention was excellent.

13. The other considerations of method, concerning propositions, are chiefly touching the utmost propositions, which limit the dimensions of sciences; for every knowledge may be fitly said, besides the profundity (which is the truth and substance of it, that makes it solid), to have a longitude and a latitude; accounting the latitude towards other sciences, and the longitude towards action; that is, from the greatest generality to the most particular precept. The one giveth rule how far one knowledge ought to intermeddle within the province of another, which is the rule they call *Καθαυτό;*[210] the other giveth rule unto what degree of particularity a knowledge should descend: which latter I find passed over in silence, being in my judgment the more material; for certainly there must be somewhat left to practice; but how much is worthy the inquiry. We see remote and superficial generalities do but offer knowledge to scorn of practical men; and are no more aiding to practice than an Ortelius'[211] universal map is to direct the way between London and York. The better sort of rules have been not unfitly compared to glasses of steel unpolished, where you may see the images of things, but first they must be filed: so the rules will help, if they be laboured and polished by practice. But how crystalline they may be made at the first, and how far forth they may be polished aforehand, is the question; the inquiry whereof seemeth to me deficient.

14.   There hath been also laboured and put in practice a method, which is not a lawful method, but a method of imposture; which is to deliver knowledges in such manner, as men may speedily come to make a show of learning who have it not: such was the travail of Raymundus Lullius, in making that art which bears his name:[212] not unlike to some books of typocosmy, which have been made since; being nothing but a mass of words of all arts, to give men countenance, that those which use the terms might be thought to understand the art; which collections are much like a fripper's or broker's shop, that hath ends of everything, but nothing of worth.

XVIII 1.   Now we descend to that part which concerneth the illustration of tradition, comprehended in that science which we call *rhetoric, or art of eloquence;* a science excellent, and excellently well laboured. For though in true value it is inferior to wisdom (as it is said by God to Moses, when he disabled himself for want of this faculty, *Aaron shall be thy speaker, and thou shalt be to him as God*),[213] yet with people it is the more mighty: so Salomon saith, *Sapiens corde appellabitur prudens, sed dulcis eloquio majora reperiet;*[214] signifying that profoundness of wisdom will help a man to a name or admiration, but that it is eloquence that prevaileth in an active life. And as to the labouring of it, the emulation of Aristotle with the rhetoricians of his time, and the experience of Cicero, hath made them in their works of rhetorics exceed themselves. Again, the excellency of examples of eloquence in the orations of Demosthenes and Cicero, added to the perfection of the precepts of eloquence, hath doubled the progression in this art; and therefore the deficiencies which I shall note will rather be in some collections, which may as handmaids attend the art, than in the rules or use of the art itself.

2.   Notwithstanding, to stir the earth a little about the roots of this science, as we have done of the rest; the duty and office of rhetoric is to apply reason to imagination for the better moving of the will. For we see reason is disturbed in the administration thereof by three means; by *illaqueation* or *sophism,* which pertains to logic; by *imagination* or *impression,* which pertains to rhetoric; and by *passion* or *affection,* which pertains to morality. And as in negotiation with others, men are wrought by cunning, by importunity, and by vehemency; so in this negotiation within ourselves, men are undermined by inconsequences, solicited and importuned by impressions or observations, and transported by passions. Neither is the nature of man so unfortunately built, as that those powers and arts should have force to disturb reason, and not to establish and advance it. For the end of logic is to teach a form of argument to secure reason, and not to entrap it; the end of morality is to procure the affections to obey reason, and not to invade it; the end of rhetoric is to fill the imagination to second reason, and not to oppress it: for these abuses of art come in but *ex obliquo,* for caution.

3.   And therefore it was great injustice in Plato, though springing out of a just hatred to the rhetoricians of his time, to esteem of rhetoric but as a voluptuary art, resembling it to cookery, that did mar wholesome meats, and help unwholesome by variety of sauces to the pleasure of the taste.[215] For we see that speech is much more conversant in adorning that which is good, than in colouring that which is evil; for there is no man but speaketh more honestly than he can do or think: and it was excellently noted by Thucydides in Cleon, that because he used to hold on the bad side in causes of estate, therefore he was ever inveighing against eloquence and good speech;[216] knowing that no man can speak fair of courses sordid and base. And therefore as Plato said elegantly, *That virtue, if she could be seen, would move great love and affection;*[217] so seeing that she cannot be showed to the sense by corporal shape, the next degree is to show her to the imagination in lively representation: for to show her to reason only in subtlety of argument, was a thing ever derided in Chrysippus and many of the Stoics; who thought to thrust virtue upon men by sharp disputations and conclusions, which have no sympathy with the will of man.

4.   Again, if the affections in themselves were pliant and obedient to reason, it were true there should be no great use of persuasions and insinuations to the will, more than of naked proposition and proofs; but in regard of the continual mutinies and seditions of the affections,

> Video meliora, proboque;
> Deteriora sequor:[218]

reason would become captive and servile, if eloquence of persuasions did not practise and win the imagination from the affections' part, and contract a confederacy between the reason and imagination against the affections; for the affections themselves carry ever an appetite to good, as reason doth. The difference is, that the affection beholdeth merely the present; reason beholdeth the future and sum of time. And therefore the present filling the imagination more, reason is commonly vanquished; but after that force of eloquence and persuasion hath made things future and remote appear as present, then upon the revolt of the imagination reason prevaileth.

5.   We conclude, therefore, that rhetoric can be no more charged with the colouring of the worse part, than logic with sophistry,[219] or morality with vice. For we know the doctrines of contraries are the same, though the use be opposite. It appeareth also that logic differeth from rhetoric, not only as the fist from the palm, the one close, the other at large; but much more in this, that logic handleth reason exact and in truth, and rhetoric handleth it as it is planted in popular opinions and manners. And therefore Aristotle[220] doth wisely place rhetoric as between logic on the one side, and moral or civil knowledge on the other, as participating of both: for the proofs and demonstrations of logic are towards all men indifferent and the same; but the proofs and persuasions of rhetoric ought to differ according to the auditors:

> Orpheus in sylvis, inter delphinas Arion.[221]

Which application, in perfection of idea, ought to extend so far, that if a man should speak of the same thing to several persons, he should speak to them all respectively and several ways: though this politic part of eloquence in private speech it is easy for the greatest orators to want: whilst by the observing their well-graced forms of speech they leese

the volubility of application: and therefore it shall not be amiss to rec-
ommend this to better inquiry, not being curious whether we place it
here, or in that part which concerneth policy.

6.   Now therefore will I descend to the deficiencies, which, as I said,
are but attendances:[222] and first, I do not find the wisdom and diligence
of Aristotle well pursued, who began to make a collection of the popular
signs and colours of good and evil, both simple and comparative, which
are as the sophisms of rhetoric, as I touched before.[223] For example:

> *Sophisma.*
> Quod laudatur, bonum: quod vituperatur, malum.
> *Redargutio.*
> Laudat venales qui vult extrudere merces.[224]

Malum est, malum est, inquit emptor: sed cum recesserit, tum glori-
abitur![225] The defects in the labour of Aristotle are three: one, that there
be but a few of many; another, that their elenches are not annexed;
and the third, that he conceiveth but a part of the use of them: for their
use is not only in probation, but much more in impression. For many
forms are equal in signification which are differing in impression; as
the difference is great in the piercing of that which is sharp and that
which is flat, though the strength of the percussion be the same: for
there is no man but will be a little more raised by hearing it said, *Your*
enemies will be glad of this:

> Hoc Ithacus velit, et magno mercentur Atridae;[226]

than by hearing it said only, *This is evil for you.*

7.   Secondly, I do resume also that which I mentioned before, touching provision or preparatory store for the furniture of speech and readiness of invention; which appeareth to be of two sorts; the one in resemblance to a shop of pieces unmade up, the other to a shop of things
ready made up; both to be applied to that which is frequent and most
in request: the former of these I will call *antitheta,* and the latter *formulae.*

8.   *Antitheta* are *theses* argued *pro et contra;* wherein men may be
more large and laborious: but, in such as are able to do it, to avoid
polixity of entry, I wish the seeds of the several arguments to be cast
up into some brief and acute sentences, not to be cited, but to be as

skeins or bottoms of thread, to be unwinded at large when they come to be used; supplying authorities and examples by reference.

> *Pro verbis legis.*
> Non est interpretatio, sed divinatio, quae recedit a litera:
> Cum receditur a litera, judex transit in legislatorem.

> *Pro sententia legis.*
> Ex omnibus verbis est eliciendus sensus qui interpretatur singula.

9.   *Formulae* are but decent and apt passages or conveyances of speech, which may serve indifferently for differing subjects; as of *preface, conclusion, digression, transition, excusation, etc.* For as in buildings, there is great pleasure and use in the well casting of the staircases, entries, doors, windows, and the like; so in speech, the conveyances and passages are of special ornament and effect.

> *A conclusion in a deliberative.*
> So may we redeem the faults passed, and prevent the inconveniences future.

XIX   1.   There remain two appendices touching the tradition of knowledge, the one critical, the other pedantical. For all knowledge is either delivered by teachers, or attained by men's proper endeavours: and therefore as the principal part of tradition of knowledge concerneth chiefly writing of books, so the relative part thereof concerneth reading of books; whereunto appertain incidently these considerations. The first is concerning the true correction and edition of authors; wherein nevertheless rash diligence hath done great prejudice. For these critics have often presumed, that that which they understand not is false set down: as the priest that, where he found it written of Saint Paul, *Demissus est per sportam,*[227] mended his book and made it *Demissus est per portam;* because *sporta* was a hard word, and out of his reading: and surely their errors, though they be not so palpable and ridiculous, yet are of the same kind. And therefore, as it hath been wisely noted, the most corrected copies are commonly the least correct.

The second is concerning the exposition and explication of authors, which resteth in annotations and commentaries: wherein it is over usual to blanch the obscure places, and discourse upon the plain.

The third is concerning the times, which in many cases give great light to true interpretations.

The fourth is concerning some brief censure and judgment of the authors; that men thereby may make some election unto themselves what books to read.

And the fifth is concerning the syntax and disposition of studies; that men may know in what order or pursuit to read.

2. For pedantical knowledge, it containeth that difference of tradition which is proper for youth; whereunto appertain divers considerations of great fruit.

As first, the timing and seasoning of knowledges; as with what to initiate them, and from what for a time to refrain them.

Secondly, the consideration where to begin with the easiest, and so proceed to the more difficult; and in what courses to press the more difficult, and then to turn them to the more easy: for it is one method to practise swimming with bladders, and another to practise dancing with heavy shoes.

A third is the application of learning according unto the propriety of the wits; for there is no defect in the faculties intellectual, but seemeth to have a proper cure contained in some studies; as, for example, if a child be bird-witted, that is, hath not the faculty of attention, the mathematics giveth a remedy thereunto; for in them, if the wit be caught away but a moment, one is new to begin. And as sciences have a propriety towards faculties for cure and help, so faculties or powers have a sympathy towards sciences for excellency or speedy profiting: and therefore it is an inquiry of great wisdom, what kinds of wits and natures are most apt and proper for what sciences.

Fourthly, the ordering of exercises is matter of great consequence to hurt or help: for, as is well observed by Cicero,[228] men in exercising their faculties, if they be not well advised, do exercise their faults and get ill habits as well as good; so there is a great judgment to be had in the continuance and intermission of exercises. It were too long to particularize a number of other considerations of this nature, things but of mean appearance, but of singular efficacy. For as the wronging or cherishing of seeds or young plants is that that is most important to their thriving (and as it was noted that the first six kings being in truth as tutors of the state of Rome in the infancy thereof, was the principal cause of the immense greatness of that state which followed), so the culture and manurance of minds in youth, hath such a forcible, though unseen operation, as hardly any length of time or contention of labour can countervail it afterwards. And it is not amiss to observe also how small and mean faculties gotten by education, yet when they fall into great men or great matters, do work great and important effects; whereof we see a notable example in Tacitus[229] of two stage players, Percennius and Vibulenus, who by their faculty of playing put the Pannonian armies into an extreme tumult and combustion. For there arising a mutiny amongst them upon the death of Augustus Caesar, Blaesus the lieutenant had committed some of the mutineers, which were suddenly rescued; whereupon Vibulenus got to be heard speak, which he did in this manner: *These poor innocent wretches appointed to cruel death, you have restored to behold the light; but who shall restore my brother to me, or life unto my brother, that was sent hither in message from the legions of Germany, to treat of the common cause? and he hath murdered him this last night by some of his fencers and ruffians, that he hath about him for his executioners upon soldiers. Answer, Blaesus, what is done with his body? The mortalest enemies do not deny burial. When I have performed my last duty to the corpse with kisses, with tears, command me to be slain beside him; so that these my fellows, for our good meaning and our true hearts to the legions, may have leave to bury us.* With which speech he put the army into an infinite fury and uproar: whereas truth was he had no brother, neither was there any such matter; but he played it merely as if he had been upon the stage.

3. But to return: we are now come to a period of rational knowledges; wherein if I have made the divisions other than those that are received, yet would I not be thought to disallow all those divisions which I do not use. For there is a double necessity imposed upon me of altering the divisions. The one, because it differeth in end and purpose, to sort together those things which are next in nature, and those things which are next in use. For if a secretary of estate should sort his papers, it is

like in his study or general cabinet he would sort together things of a nature, as treaties, instructions, etc., but in his boxes or particular cabinet he would sort together those that he were like to use together, though of several natures; so in this general cabinet of knowledge it was necessary for me to follow the divisions of the nature of things; whereas if myself had been to handle any particular knowledge, I would have respected the divisions fittest for use. The other, because the bringing in of the deficiencies did by consequence alter the partitions of the rest. For let the knowledge extant, for demonstration sake, be fifteen; let the knowledge with the deficiencies be twenty; the parts of fifteen are not the parts of twenty; for the parts of fifteen are three and five; the parts of twenty are two, four, five, and ten. So as these things are without contradiction, and could not otherwise be.

XX 1. We proceed now to that knowledge which considereth of the appetite and will of man: whereof Salomon saith, *Ante omnia, fili, custodi cor tuum; nam inde procedunt actiones vitae.*[230] In the handling of this science, those which have written seem to me to have done as if a man, that professed to teach to write, did only exhibit fair copies of alphabets and letters joined, without giving any precepts or directions for the carriage of the hand and framing of the letters. So have they made good and fair exemplars and copies, carrying the draughts and portraitures of *good, virtue, duty, felicity;* propounding them well described as the true objects and scopes of man's will and desires. But how to attain these excellent marks, and how to frame and subdue the will of man to become true and conformable to these pursuits, they pass it over altogether, or slightly and unprofitably. For it is not the disputing that moral virtues are in the mind of man by habit and not by nature,[231] or the distinguishing that generous spirits are won by doctrines and persuasions, and the vulgar sort by reward and punishment, and the like scattered glances and touches, that can excuse the absence of this part.

2. The reason of this omission I suppose to be that hidden rock whereupon both this and many other barks of knowledge have been cast away; which is, that men have despised to be conversant in ordi-

nary and common matters, the judicious direction whereof nevertheless is the wisest doctrine (for life consisteth not in novelties or
subtleties), but contrariwise they have compounded sciences chiefly
of a certain resplendent or lustrous mass of matter, chosen to give glory
either to the subtlety of disputations, or to the eloquence of discourses.
But Seneca giveth an excellent check to eloquence: *Nocet illis eloquentia,
quibus non rerum cupiditatem facit, sed sui.*[232] Doctrine should be such
as should make men in love with the lesson, and not with the teacher;
being directed to the auditor's benefit, and not to the author's commendation. And therefore those are of the right kind which may be
concluded as Demosthenes concludes his counsel, *Quae si feceritis, non
oratorem duntaxat in praesentia laudabitis, sed vosmetipsos etiam non
ita multo post statu rerum vestrarum meliore.*[233]

3.   Neither needed men of so excellent parts to have despaired of a
fortune, which the poet Virgil promised himself, and indeed obtained,
who got as much glory of eloquence, wit, and learning in the expressing
of the observations of husbandry, as of the heroical acts of Aeneas:—

> Nec sum animi dubius, verbis ea vincere magnum
> Quam sit, et angustis his addere rebus honorem.[234]

And surely, if the purpose be in good earnest, not to write at leisure
that which men may read at leisure, but really to instruct and suborn
action and active life, these Georgics of the mind, concerning the husbandry and tillage thereof, are no less worthy than the heroical descriptions of virtue, duty, and felicity. Wherefore the main and primitive
division of moral knowledge seemeth to be into the exemplar or platform of good, and the regiment or culture of the mind: the one describing the nature of good, the other prescribing rules how to subdue,
apply, and accommodate the will of man thereunto.

4.   The doctrine touching the platform or nature of good considereth it either simple or compared; either the kinds of good, or the degrees of good; in the latter whereof those infinite disputations, which
were touching the supreme degree thereof, which they term felicity,
beatitude, or the highest good, the doctrines concerning which were
as the heathen divinity,[235] are by the Christian faith discharged. And as
Aristotle saith, *That young men may be happy, but not otherwise but by*

hope;[236] so we must all acknowledge our minority, and embrace the felicity which is by hope of the future world.

5.   Freed therefore and delivered from this doctrine of the philosopher's heaven, whereby they feigned a higher elevation of man's nature than was (for we see in what a height of style Seneca writeth, *Vere magnum, habere fragilitatem hominis, securitatem Dei*[237]), we may with more sobriety and truth receive the rest of their inquiries and labours. Wherein for the nature of good positive or simple, they have set it down excellently, in describing the forms of virtue and duty, with their situations and postures; in distributing them into their kinds, parts, provinces, actions, and administrations, and the like: nay further, they have commended them to man's nature and spirit, with great quickness of argument and beauty of persuasions; yea, and fortified and entrenched them, as much as discourse can do, against corrupt and popular opinions. Again, for the degrees and comparative nature of good, they have also excellently handled it in their triplicity of good, in the comparison between a contemplative and an active life,[238] in the distinction between virtue with reluctation and virtue secured, in their encounters between honesty and profit, in their balancing of virtue with virtue, and the like; so as this part deserveth to be reported for excellently laboured.

6.   Notwithstanding, if before they had come to the popular and received notions of virtue and vice, pleasure and pain, and the rest, they had stayed a little longer upon the inquiry concerning the roots of good and evil, and the strings of those roots, they had given, in my opinion, a great light to that which followed; and specially if they had consulted with nature, they had made their doctrines less prolix and more profound; which being by them in part omitted and in part handled with much confusion, we will endeavour to resume and open in a more clear manner.

7.   There is formed in every thing a double nature of good: the one, as every thing is a total or substantive in itself; the other, as it is a part or member of a greater body; whereof the latter is in degree the greater and the worthier, because it tendeth to the conservation of a more general form. Therefore we see the iron in particular sympathy moveth to the lodestone; but yet if it exceed a certain quantity, it forsaketh the affection to the lodestone, and like a good patriot moveth to the earth,

which is the region and country of massy bodies: so may we go forward, and see that water and massy bodies move to the centre of the earth; but rather than to suffer a divulsion in the continuance of nature, they will move upwards from the centre of the earth, forsaking their duty to the earth in regard to their duty to the world. This double nature of good, and the comparative thereof, is much more engraven upon man, if he degenerate not: unto whom the conservation of duty to the public ought to be much more precious than the conservation of life and being: according to that memorable speech of Pompeius Magnus, when being in commission of purveyance for a famine at Rome, and being dissuaded with great vehemency and instance by his friends about him that he should not hazard himself to sea in an extremity of weather, he said only to them, *Necesse est ut eam, non ut vivam.*[239] But it may be truly affirmed that there was never any philosophy, religion, or other discipline, which did so plainly and highly exalt the good which is communicative, and depress the good which is private and particular, as the Holy Faith; well declaring that it was the same God that gave the Christian law to men, who gave those laws of nature to inanimate creatures that we spoke of before; for we read that the elected saints of God have wished themselves anathematized and razed out of the book of life, in an ecstasy of charity and infinite feeling of communion.[240]

8. This being set down and strongly planted, doth judge and determine most of the controversies wherein moral philosophy is conversant. For first, it decideth the question touching the preferment of the contemplative or active life, and decideth it against Aristotle. For all the reasons which he bringeth for the contemplative are private, and respecting the pleasure and dignity of a man's self (in which respects, no question, the contemplative life hath the preeminence), not much unlike to that comparison, which Pythagoras made for the gracing and magnifying of philosophy and contemplation: who being asked what he was, answered, *That if Hiero were ever at the Olympian games, he knew the manner, that some came to try their fortune for the prizes, and some came as merchants to utter their commodities, and some came to make good cheer and meet their friends, and some came to look on; and that he was one of them that came to look on.*[241] But men must know, that in this theatre of man's life it is reserved only for God and angels to be lookers on: neither could the like question ever have been received in the church (notwithstanding their *Pretiosa in oculis Domini mors sanctorum ejus,*[242] by which place they would exalt their civil death and regular professions), but upon this defence, that the monastical life is not simple contemplative, but performeth the duty either of incessant prayers and supplications, which hath been truly esteemed as an office in the church, or else of writing or taking instructions for writing concerning the law of God, as Moses did when he abode so long in the mount.[243] And so we see Enoch the seventh from Adam, who was the first contemplative and walked with God, yet did also endow the church with prophecy, which Saint Jude citeth.[244] But for contemplation which should be finished in itself, without casting beams upon society, assuredly divinity knoweth it not.

9.   It decideth also the controversies between Zeno and Socrates, and their schools and successions, on the one side, who placed felicity in virtue simply or attended, the actions and exercises whereof do chiefly embrace and concern society; and on the other side, the Cyrenaics and Epicureans, who placed it in pleasure, and made virtue (as it is used in some comedies of errors, wherein the mistress and the maid change habits) to be but as a servant, without which pleasure cannot be served and attended, and the reformed school of the Epicureans, which placed it in serenity of mind and freedom from perturbation (as if they would have deposed Jupiter again, and restored Saturn and the first age, when there was no summer nor winter, spring nor autumn, but all after one air and season), and Herillus, who placed felicity in extinguishment of the disputes of the mind, making no fixed nature of good and evil, esteeming things according to the clearness of the desires, or the reluctation; which opinion was revived in the heresy of the Anabaptists,[245] measuring things according to the motions of the spirit, and the constancy or wavering of belief: all which are manifest to tend to private repose and contentment, and not to point of society.

10.   It censureth also the philosophy of Epictetus, which presupposeth that felicity must be placed in those things which are in our power, lest we be liable to fortune and disturbance: as if it were not a thing much more happy to fail in good and virtuous ends for the public,

than to obtain all that we can wish to ourselves in our proper fortune; as Consalvo said to his soldiers, showing them Naples, and protesting, *He had rather die one foot forwards, than to have his life secured for long by one foot of retreat.*[246] Whereunto the wisdom of that heavenly leader hath signed, who hath affirmed that *a good conscience is a continual feast;*[247] showing plainly that the conscience of good intensions, howsoever succeeding, is a more continual joy to nature, than all the provision which can be made for security and repose.

11.   It censureth likewise that abuse of philosophy, which grew general about the time of Epictetus, in converting it into an occupation or profession; as if the purpose had been, not to resist and extinguish perturbations, but to fly and avoid the causes of them, and to shape a particular kind and course of life to that end; introducing such a health of mind, as was that health of body of which Aristotle speaketh of Herodicus, who did nothing all his life long but intend his health:[248] whereas if men refer themselves to duties of society, as that health of body is best, which is ablest to endure all alterations and extremities; so likewise that health of mind is most proper, which can go through the greatest temptations and perturbations. So as Diogenes' opinion is to be accepted, who commended not them which abstained, but them which sustained, and could refrain their mind *in praecipitio*, and could give unto the mind, as is used in horsemanship, the shortest stop or turn.[249]

12.   Lastly, it censureth the tenderness and want of application in some of the most ancient and reverend philosophers and philosophical men, that did retire too easily from civil business, for avoiding of indignities and perturbations: whereas the resolution of men truly moral ought to be such as the same Consalvo said the honour of a soldier should be, *e tela crassiore*, and not so fine as that everything should catch in it and endanger it.

XXI

1.   To resume *private* or *particular good;* it falleth into the division of *good active and passive:* for this difference of good, not unlike to that which amongst the Romans was expressed in the familiar or household terms of *promus* and *condus,* is formed also in all things, and is best disclosed in the two several appetites in creatures;

the one to preserve or continue themselves, and the other to dilate or multiply themselves; whereof the latter seemeth to be the worthier: for in nature the heavens, which are the more worthy, are the agent; and the earth, which is the less worthy, is the patient. In the pleasures of living creatures, that of generation is greater than that of food; in divine doctrine, *beatius est dare quam accipere,*[250] and in life, there is no man's spirit so soft, but esteemeth the effecting of somewhat that he hath fixed in his desire, more than sensuality; which priority of the active good, is much upheld by the consideration of our estate to be mortal and exposed to fortune. For if we might have a perpetuity and certainty in our pleasures, the state of them would advance their price: but when we see it is but *magni aestimamus mori tardius,*[251] and *ne glorieris de crastino, nescis partum diei,*[252] it maketh us to desire to have somewhat secured and exempted from time, which are only our deeds and works: as it is said, *opera eorum sequuntur eos.*[253] The preeminence likewise of this active good is upheld by the affection which is natural in man towards variety and proceeding; which in the pleasures of the sense, which is the principal part of passive good, can have no great latitude: *Cogita quamdiu eadem feceris; cibus, somnus, ludus; per hunc circulum curritur; mori velle non tantum fortis, aut miser, aut prudens, sed etiam fastidiosus potest.*[254] But in enterprises, pursuits, and purposes of life, there is much variety; whereof men are sensible with pleasure in their inceptions, progressions, recoils, reintegrations, approaches and attainings to their ends: so as it was well said, *Vita sine proposito languida et vaga est.*[255] Neither hath this active good any identity with the good of society, though in some case it hath an incidence into it; for although it do many times bring forth acts of beneficence, yet it is with a respect private to a man's own power, glory, amplification, continuance; as appeareth plainly, when it findeth a contrary subject. For that gigantine state of mind which possesseth the troublers of the world, such as was Lucius Sulla, and infinite other in smaller model, who would have all men happy or unhappy as they were their friends or enemies, and would give form to the world, according to their own humours (which is the true Theomachy), pretendeth and aspireth to active good, though it recedeth furthest from good of society, which we have determined to be the greater.

2. To resume passive good, it receiveth a subdivision of conservative and perfective. For let us take a brief review of that which we have said: we have spoken first of the good of society, the intention whereof embraceth the form of human nature, whereof we are members and portions, and not our own proper and individual form: we have spoken of active good, and supposed it as a part of private and particular good: and rightly, for there is impressed upon all things a triple desire or appetite proceeding from love to themselves; one of preserving and continuing their form; another of advancing and perfecting their form; and a third of multiplying and extending their form upon other things; whereof the multiplying, or signature of it upon other things, is that which we handled by the name of active good. So as there remaineth the conserving of it, and perfecting or raising of it; which latter is the highest degree of passive good. For to preserve in state is the less, to preserve with advancement is the greater. So in man,

Igneus est ollis vigor, et caelestis origo.[256]

His approach or assumption to divine or angelical nature is the perfection of his form; the error or false imitation of which good is that which is the tempest of human life; while man, upon the instinct of an advancement formal and essential, is carried to seek an advancement local. For as those which are sick, and find no remedy, do tumble up and down and change place, as if by a remove local they could obtain a remove internal; so is it with men in ambition, when failing of the means to exalt their nature, they are in a perpetual estuation to exalt their place. So then passive good is, as was said, either conservative or perfective.

3. To resume the good of conservation or comfort, which consisteth in the fruition of that which is agreeable to our natures; it seemeth to be the most pure and natural of pleasures, but yet the softest and the lowest. And this also receiveth a difference, which hath neither been well judged of, nor well inquired: for the good of fruition or contentment is placed either in the sincereness of the fruition, or in the quickness and vigour of it; the one superinduced by equality, the other by vicissitude; the one having less mixture of evil, the other more impression of good. Which of these is the greater good is a question controverted; but whether man's nature may not be capable of both, is a question not inquired.

4.   The former question being debated between Socrates and a sophist, Socrates placing felicity in an equal and constant peace of mind, and the sophist in much desiring and much enjoying, they fell from argument to ill words: the sophist saying that Socrates' felicity was the felicity of a block or stone; and Socrates saying that the sophist's felicity was the felicity of one that had the itch, who did nothing but itch and scratch.[257] And both these opinions do not want their supports. For the opinion of Socrates is much upheld by the general consent even of the Epicures themselves, that virtue beareth a great part in felicity; and if so, certain it is, that virtue hath more use in clearing perturbations than in compassing desires. The sophist's opinion is much favoured by the assertion we last spoke of, that good of advancement is greater than good of simple preservation; because every obtaining a desire hath a show of advancement, as motion though in a circle has a show of progression.

5.   But the second question, decided the true way, maketh the former superfluous. For can it be doubted but that there are some who take more pleasure in enjoying pleasures than some other, and yet nevertheless are less troubled with the loss or leaving of them? So as this same, *Non uti ut non appetas, non appetere ut non metuas, sunt animi pusilli et diffidentis.* And it seemeth to me, that most of the doctrines of the philosophers are more fearful and cautionary than the nature of things requireth. So have they increased the fear of death in offering to cure it. For when they would have a man's whole life to be but a discipline or preparation to die, they must needs make men think that it is a terrible enemy, against whom there is no end of preparing. Better saith the poet:

> Qui spatium vitae extremum inter munera ponat
> Naturae.[258]

So have they sought to make men's minds too uniform and harmonical, by not breaking them sufficiently to contrary motions: the reason whereof I suppose to be, because they themselves were men dedicated to a private, free, and unapplied course of life. For as we see, upon the lute or like instrument, a ground, though it be sweet and have show of many changes, yet breaketh not the hand to such strange and hard stops and passages as a set song or voluntary; much after the same manner was the diversity between a philosophical and a civil life. And therefore men are to imitate the wisdom of jewellers; who, if there be a grain, or a cloud, or an ice,[259] which may be ground forth without taking too much of the stone, they help it; but if it should lessen and abate the stone too much, they will not meddle with it: so ought men so to procure serenity as they destroy not magnanimity.

6.   Having therefore deduced the good of man which is private and particular, as far as seemeth fit; we will now return to that good of man which respecteth and beholdeth society, which we may term Duty; because the term of Duty is more proper to a mind well framed and disposed towards others, as the term of virtue is applied to a mind well formed and composed in itself: though neither can a man understand virtue without some relation to society, nor Duty without an inward disposition. This part may seem at first to pertain to science civil and politic: but not if it be well observed; for it concerneth the regiment and government of every man over himself, and not over others. And as in architecture the direction of framing the posts, beams, and other parts of building, is not the same with the manner of joining them and erecting the building; and in mechanicals, the direction how to frame an instrument or engine, is not the same with the manner of setting it on work and employing it (and yet nevertheless in expressing of the one you incidentally express the aptness towards the other); so the doctrine of conjugation of men in society differeth from that of their conformity thereunto.

7.   This part of Duty is subdivided into two parts: the common Duty of every man, as a man or member of a state; the other, the respective or special Duty of every man, in his profession, vocation, and place. The first of these is extant and well laboured, as hath been said. The second likewise I may report rather dispersed than deficient; which manner of dispersed writing in this kind of argument I acknowledge to be best. For who can take upon him to write of the proper duty, virtue, challenge, and right of every several vocation, profession, and place? For although sometimes a looker on may see more than a

gamester, and there be a proverb more arrogant than sound, *that the vale best discovereth the hill;* yet there is small doubt but that men can write best, and most really and materially, in their own professions; and that the writing of speculative men of active matter, for the most part, doth seem to men of experience as Phormio's argument of the wars seemed to Hannibal, to be but dreams and dotage.[260] Only there is one vice which accompanieth them that write in their own professions, that they magnify them in excess. But generally it were to be wished, as that which would make learning indeed solid and fruitful, that active men would or could become writers.

8. In which kind I cannot but mention, *honoris causa,* your Majesty's excellent book touching the duty of a king;[261] a work richly compounded of divinity, morality, and policy, with great aspersion of all other arts; and being, in mine opinion, one of the most sound and healthful writings that I have read; not distempered in the heat of invention, nor in the coldness of negligence; not sick of dizziness, as those are who leese themselves in their order; nor of convulsions, as those which cramp in matters impertinent; not savouring of perfumes and paintings, as those do who seek to please the reader more than nature beareth; and chiefly well disposed in the spirits thereof, being agreeable to truth and apt for action; and far removed from that natural infirmity, whereunto I noted those that write in their own professions to be subject, which is, that they exalt it above measure: for your Majesty hath truly described, not a king of Assyria or Persia in their extern glory, but a Moses or a David, pastors of their people. Neither can I ever leese out of my remembrance, what I heard your Majesty, in the same sacred spirit of Government, deliver in a great cause of judicature, which was, *That kings ruled by their laws as God did by the laws of nature; and ought as rarely to put in use their supreme prerogative, as God doth his power of working miracles.* And yet notwithstanding, in your book of a free monarchy,[262] you do well give men to understand that you know the plenitude of the power and right of a king, as well as the circle of his office and duty. Thus have I presumed to allege this excellent writing of your Majesty, as a prime or eminent example of tractates concerning special and respective duties: wherein I should have said as much, if it had been written a thousand years since: neither am I moved with certain courtly decencies, which esteem it flattery to praise in presence; no, it is flattery to praise in absence; that is, when either the virtue is absent, or the occasion is absent; and so the praise is not natural, but forced, either in truth or in time. But let Cicero be read in his oration *pro Marcello*, which is nothing but an excellent table of Caesar's virtue, and made to his face; besides the example of many other excellent persons, wiser a great deal than such observers; and we will never doubt, upon a full occasion, to give just praises to present or absent.

9.   But to return: there belongeth further to the handling of this part, touching the duties of professions and vocations, a relative or opposite, touching the frauds, cautels, impostures, and vices of every profession, which hath been likewise handled: but how? rather in a satire and cynically than seriously and wisely: for men have rather sought by wit to deride and traduce much of that which is good in professions, than with judgment to discover and sever that which is corrupt. For, as Salomon saith, he that cometh to seek after knowledge with a mind to scorn and censure, shall be sure to find matter for his humour, but no matter for his instruction: *Quaerenti derisori scientiam ipsa se abscondit; sed studioso fit obviam.*[263] But the managing of this argument with integrity and truth, which I note as deficient, seemeth to me to be one of the best fortifications for honesty and virtue that can be planted. For, as the fable goeth of the basilisk, that if he see you first, you die for it; but if you see him first, he dieth: so it is with deceits and evil arts; which, if they be first espied they leese their life; but if they prevent, they endanger. So that we are much beholden to Machiavel and others, that write what men do, and not what they ought to do. For it is not possible to join serpentine wisdom with columbine innocency,[264] except men know exactly all the conditions of the serpent: his baseness and going upon his belly, his volubility and lubricity, his envy and sting, and the rest; that is, all forms and natures of evil: for without this, virtue lieth open and unfenced. Nay, an honest man can do no good upon those that are wicked to reclaim them, without the help of the knowledge of evil. For men of corrupted minds presuppose that honesty groweth out of simplicity of manners, and believing of preachers, schoolmasters, and men's exterior language: so as, except you can make them perceive that you know the utmost reaches of their own

corrupt opinions, they despise all morality; *Non recipit stultus verba prudentiae, nisi ea dixeris quae versantur in corde ejus.*[265]

10.   Unto this part, touching Respective Duty, doth also appertain the duties between husband and wife, parent and child, master and servant: so likewise the laws of friendship and gratitude, the civil bond of companies, colleges, and politic bodies, of neighbourhood, and all other proportionate duties; not as they are parts of government and society, but as to the framing of the mind of particular persons.

11.   The knowledge concerning good respecting Society doth handle it also, not simply alone, but comparatively; whereunto belongeth the weighing of duties between person and person, case and case, particular and public: as we see in the proceeding of Lucius Brutus against his own sons, which was so much extolled; yet what was said?

Infelix, utcunque ferent ea facta minores.[266]

So the case was doubtful, and had opinion on both sides. Again, we see when M. Brutus and Cassius invited to a supper certain whose opinions they meant to feel, whether they were fit to be made their associates, and cast forth the question touching the killing of a tyrant being a usurper, they were divided in opinion;[267] some holding that servitude was the extreme of evils, and others that tyranny was better than a civil war: and a number of the like cases there are of comparative duty; amongst which that of all others is the most frequent, where the question is of a great deal of good to ensue of a small injustice. Which Jason of Thessalia determined against the truth: *Aliqua sunt injuste facienda, ut multa juste fieri possint.*[268] But the reply is good, *Auctorem praesentis justitiae habes, sponsorem futurae non habes.* Men must pursue things which are just in present, and leave the future to the divine Providence. So then we pass on from this general part touching the exemplar and description of good.

XXII   **1.**   Now therefore that we have spoken of this fruit of life, it remaineth to speak of the husbandry that belongeth thereunto; without which part the former seemeth to be no better than a fair image, or statua, which is beautiful to contemplate, but is without life

Statue without life or motion

and motion; whereunto Aristotle himself subscribeth in these words: *Necesse est scilicet de virtute dicere, et quid sit, et ex quibus gignatur. Inutile enim fere fuerit virtutem quidem nosse, acquirendae autem ejus modos et vias ignorare: non enim de virtute tantum, qua specie sit, quaerendum est, sed et quomodo sui copiam faciat: utrumque enim volumus, et rem ipsam nosse, et ejus compotes fieri: hoc autem ex voto non succedet, nisi sciamus et ex quibus et quomodo.*[269] In such full words and with such iteration doth he inculcate this part. So saith Cicero in great commendation of Cato the second, that he had applied himself to philosophy, *Non ita disputandi causa, sed ita vivendi.*[270] And although the neglect of our times, wherein few men do hold any consultations touching the reformation of their life (as Seneca excellently saith, *De partibus vitae quisque deliberat, de summa nemo*),[271] may make this part seem superfluous; yet I must conclude with that aphorism of Hippocrates, *Qui gravi morbo correpti dolores non sentiunt, iis mens aegrotat,*[272] they need medicine, not only to assuage the disease, but to awake the sense. And if it be said, that the cure of men's minds belongeth to sacred divinity, it is most true: but yet moral philosophy may be preferred unto her as a wise servant and humble handmaid. For as the Psalm saith, *That the eyes of the handmaid look perpetually towards the mistress,*[273] and yet no doubt many things are left to the discretion of the handmaid, to discern of the mistress's will; so ought moral philosophy to give a constant attention to the doctrines of divinity, and yet so as it may yield of herself, within due limits, many sound and profitable directions.

2.   This part therefore, because of the excellency thereof, I cannot but find exceeding strange that it is not reduced to written inquiry: the rather, because it consisteth of much matter, wherein both speech and action is often conversant; and such wherein the common talk of men (which is rare, but yet cometh sometimes to pass) is wiser than their books. It is reasonable therefore that we propound it in the more particularity, both for the worthiness, and because we may acquit ourselves for reporting it deficient; which seemeth almost incredible, and is otherwise conceived and presupposed by those themselves that have written. We will therefore enumerate some heads or points thereof, that it may appear the better what it is, and whether it be extant.

156

3.   First therefore in this, as in all things which are practical, we ought to cast up our account, what is in our power, and what not; for the one may be dealt with by way of alteration, but the other by way of application only. The husbandman cannot command neither the nature of the earth nor the seasons of the weather; no more can the physician the constitution of the patient, nor the variety of accidents. So in the culture and cure of the mind of man, two things are without our command; points of nature, and points of fortune. For to the basis of the one, and the conditions of the other, our work is limited and tied. In these things therefore it is left unto us to proceed by application;

   Vincenda est omnis fortuna ferendo:[274]

and so likewise,

   Vincenda est omnis Natura ferendo.

But when that we speak of suffering, we do not speak of a dull and neglected suffering, but of a wise and industrious suffering, which draweth and contriveth use and advantage out of that which seemeth adverse and contrary; which is that property which we call accommodating or applying. Now the wisdom of application resteth principally in the exact and distinct knowledge of the precedent state or disposition, unto which we do apply: for we cannot fit a garment, except we first take measure of the body.

4.   So then the first article of this knowledge is to set down sound and true distributions and descriptions of the several characters and tempers of men's natures and dispositions; especially having regard to those differences which are most radical in being the fountains and causes of the rest, or most frequent in concurrence or commixture; wherein it is not the handling of a few of them in passage, the better to describe the mediocrities of virtues, that can satisfy this intention. For if it deserve to be considered, that there are minds which are proportioned to great matters, and others to small[275] (which Aristotle handleth, or ought to have handled, by the name of magnanimity), doth it not deserve as well to be considered, that there are minds proportioned to intend many matters, and others to few? So that some can divide themselves: others can perchance do exactly well, but it must be in few things

at once: and so there cometh to be a narrowness of mind, as well as a pusil-lanimity. And again, that some minds are proportioned to that which may be dispatched at once, or within a short return of time; others to that which begins afar off, and is to be won with length of pursuit:

Jam tum tenditque fovetque.[276] *Such already was her aim and purpose*

So that there may be fitly said to be a longanimity, which is commonly also ascribed to God as a magnanimity. So further deserved it to be considered by Aristotle, *that there is a disposition in conversation (sup-posing it in things which do in no sort touch or concern a man's self) to soothe and please; and a disposition contrary to contradict and cross:* and deserveth it not much better to be considered, *that there is a disposition, not in conversation or talk, but in matter of more serious nature (and supposing it still in things merely indifferent), to take pleasure in the good of another: and a disposition contrariwise, to take distaste at the good of another?*[277] which is that properly which we call good nature or ill na-ture, benignity or malignity: and therefore I cannot sufficiently marvel that this part of knowledge, touching the several characters of natures and dispositions, should be omitted both in morality and policy; con-sidering it is of so great ministry and suppeditation to them both. A man shall find in the traditions of astrology some pretty and apt di-visions of men's natures, according to the predominances of the plan-ets; *lovers of quiet, lovers of action, lovers of victory, lovers of honour, lovers of pleasure, lovers of arts, lovers of change,* and so forth. A man shall find in the wisest sort of these relations which the Italians make touching conclaves, the natures of the several cardinals handsomely and lively painted forth: a man shall meet with in every day's conference the denominations of *sensitive, dry, formal, real, humorous, certain, huomo di prima impressione, huomo di ultima impressione,* and the like: and yet nevertheless this kind of observation wandereth in words, but is not fixed in inquiry. For the distinctions are found, many of them, but we conclude no precepts upon them: wherein our fault is the greater; because both history, poesy, and daily experience are as goodly fields where these observations grow; whereof we make a few posies to hold in our hands, but no man bringeth them to the confectionary, that re-ceipts might be made of them for use of life.

Nic. Ethics 4.6

5.   Of much like kind are those impressions of nature, which are imposed upon the mind by the sex, by the age, by the region, by health
and sickness, by beauty and deformity, and the like, which are inherent
and not extern; and again, those which are caused by extern fortune;
as sovereignty, nobility, obscure birth, riches, want, magistracy, privateness, prosperity, adversity, constant fortune, variable fortune, rising
per saltum, per gradus, and the like. And therefore we see that Plautus
maketh it a wonder to see an old man beneficent, *benignitas hujus ut
adolescentuli est.*[278] Saint Paul concludeth that severity of discipline was
to be used to the Cretans, *increpa eos dure,* upon the disposition of
their country, *Cretenses semper mendaces, malae bestiae, ventres pigri.*[279]
Sallust noteth that it is usual with kings to desire contradictories: *Sed
plerumque regiae voluntates, ut vehementes sunt, sic mobiles, saepeque
ipsae sibi adversae.*[280] Tacitus observeth how rarely raising of the fortune mendeth the disposition: *solus Vespasianus mutatus in melius.*[281]
Pindarus maketh an observation, that great and sudden fortune for
the most part defeateth men *qui magnam felicitatem concoquere non
possunt.*[282] So the Psalm showeth it is more easy to keep a measure in
the enjoying of fortune, than in the increase of fortune: *divitiae si
affluant, nolite cor apponere.*[283] These observations, and the like, I deny
not but are touched a little by Aristotle, as in passage in his Rhetorics,[284]
and are handled in some scattered discourses: but they were never in-
corporated into moral philosophy, to which they do essentially appertain; as the knowledge of the diversity of grounds and moulds doth
to agriculture, and the knowledge of the diversity of complexions and
constitutions doth to the physician; except we mean to follow the indis-
cretion of empirics, which minister the same medicines to all patients.
6.   Another article of this knowledge is the inquiry touching the
affections; for as in medicining of the body, it is in order first to know
the divers complexions and constitutions; secondly, the diseases; and
lastly, the cures: so in medicining of the mind, after knowledge of the
divers characters of men's natures, it followeth, in order, to know the
diseases and infirmities of the mind, which are no other than the perturbations and distempers of the affections. For as the ancient politiques
in popular estates[285] were wont to compare the people to the sea, and
the orators to the winds; because as the sea would of itself be calm and

159

quiet, if the winds did not move and trouble it; so the people would be peaceable and tractable, if the seditious orators did not set them in working and agitation: so it may be fitly said, that the mind in the nature thereof would be temperate and stayed, if the affections, as winds, did not put it into tumult and perturbation. And here again I find strange, as before, that Aristotle should have written divers volumes of ethics, and never handled the affections, which is the principal subject thereof; and yet in his Rhetorics, where they are considered but collaterally, and in a second degree, as they may be moved by speech, he findeth place for them,[286] and handleth them well for the quantity; but where their true place is, he pretermitteth them. For it is not his disputations about pleasure and pain that can satisfy this inquiry, no more than he that should generally handle the nature of light can be said to handle the nature of colours; for pleasure and pain are to the particular affections as light is to particular colours. Better travails, I suppose, had the Stoics taken in this argument as far as I can gather by that which we have at second hand. But yet it is like it was after their manner, rather in subtlety of definitions (which in a subject of this nature are but curiosities) than in active and ample descriptions and observations. So likewise I find some particular writings of an elegant nature, touching some of the affections; as of anger, of comfort upon adverse accidents, of tenderness of countenance, and other.[287] But the poets and writers of histories are the best doctors of this knowledge; where we may find painted forth with great life, how affections are kindled and incited; and how pacified and refrained; and how again contained from act and further degree; how they disclose themselves; how they work; how they vary; how they gather and fortify; how they are enwrapped one within another; and how they do fight and encounter one with another; and other the like particularities: amongst the which this last is of special use in moral and civil matters; how, I say, to set affection against affection, and to master one by another; even as we use to hunt beast with beast, and fly bird with bird, which otherwise percase we could not so easily recover: upon which foundation is erected that excellent use of *praemium* and *poena*, whereby civil states consist: employing the predominant affections of fear and hope, for the suppressing and bridling the rest. For as in the government of states it is sometimes necessary to bridle one faction with another, so it is in the government within.

7. Now come we to those points which are within our own command, and have force and operation upon the mind, to affect the will and appetite, and to alter manners: wherein they ought to have handled custom, exercise, habit, education, example, imitation, emulation, company, friends, praise, reproof, exhortation, fame, laws, books, studies: these as they have determinate use in moralities, from these the mind suffereth; and of these are such receipts and regiments compounded and described, as may serve to recover or preserve the health and good estate of the mind, as far as pertaineth to human medicine: of which number we will insist upon some one or two, as an example of the rest, because it were too long to prosecute all; and therefore we do resume custom and habit to speak of.

8. The opinion of Aristotle seemeth to me a negligent opinion, that of those things which consist by nature nothing can be changed by custom; using for example, that if a stone be thrown ten thousand times up, it will not learn to ascend;[288] and that by often seeing or hearing, we do not learn to see or hear the better. For though this principle be true in things wherein nature is peremptory (the reason whereof we cannot now stand to discuss), yet it is otherwise in things wherein nature admitteth a latitude. For he might see that a strait glove will come more easily on with use; and that a wand will by use bend otherwise than it grew; and that by use of the voice we speak louder and stronger; and that by use of enduring heat or cold, we endure it the better, and the like: which latter sort have a nearer resemblance unto that subject of manners he handleth, than those instances which he allegeth. But allowing his conclusion, that virtues and vices consist in habit, he ought so much the more to have taught the manner of superinducing that habit: for there be many precepts of the wise ordering the exercises of the mind, as there is of ordering the exercises of the body; whereof we will recite a few.

9. The first shall be, that we beware we take not at the first either too high a strain, or too weak: for if too high, in a diffident nature you discourage, in a confident nature you breed an opinion of facility, and so a sloth; and in all natures you breed a further expectation than can

hold out, and so an insatisfaction in the end: if too weak of the other side, you may not look to perform and overcome any great task.

10.   Another precept is, to practise all things chiefly at two several times, the one when the mind is best disposed, the other when it is worst disposed; that by the one you may gain a great step, by the other you may work out the knots and stonds of the mind, and make the middle times the more easy and pleasant.

11.   Another precept is, that which Aristotle mentioneth by the way, which is to bear ever towards the contrary extreme of that whereunto we are by nature inclined; like unto the rowing against the stream, or making a wand straight by bending[289] him contrary to his natural crookedness.[290]

12.   Another precept is, that the mind is brought to anything better, and with more sweetness and happiness, if that whereunto you pretend be not first in the intention, but *tanquam aliud agendo,* because of the natural hatred of the mind against necessity and constraint. Many other axioms there are touching the managing of exercise and custom; which being so conducted doth prove indeed another nature; but being governed by chance doth commonly prove but an ape of nature, and bringing forth that which is lame and counterfeit.

13.   So if we should handle books and studies, and what influence and operation they have upon manners, are there not divers precepts of great caution and direction appertaining thereunto? Did not one of the fathers[291] in great indignation call poesy, *vinum daemonum,* because it increaseth temptations, perturbations, and vain opinions? Is not the opinion of Aristotle worthy to be regarded, wherein he saith, *That young men are no fit auditors of moral philosophy, because they are not settled from the boiling heat of their affections, nor attempered with time and experience?*[292] And doth it not hereof come, that those excellent books and discourses of the ancient writers (whereby they have persuaded unto virtue most effectually, by representing her in state and majesty, and popular opinions against virtue in their parasites' coats fit to be scorned and derided) are of so little effect towards honesty of life, because they are not read and revolved by men in their mature and settled years, but confined almost to boys and beginners? But is it not true also, that much less young men are fit auditors of matters of

policy, till they have been thoroughly seasoned in religion and morality; lest their judgments be corrupted, and made apt to think that there are no true differences of things, but according to utility and fortune, as the verse describes it, *Prosperum et felix scelus virtus vocatur;*[293] and again, *Ille crucem pretium sceleris tulit, hic diadema:*[294] which the poets do speak satirically, and in indignation on virtue's behalf; but books of policy do speak it seriously and positively; for so it pleaseth Machiavel to say, *That if Caesar had been overthrown, he would have been more odious than ever was Catiline;*[295] as if there had been no difference but in fortune, between a very fury of lust and blood, and the most excellent spirit (his ambition reserved) of the world? Again, is there not a caution likewise to be given of the doctrines of moralities themselves (some kinds of them), lest they make men too precise, arrogant, incompatible; as Cicero saith of Cato, *In Marco Catone haec bona quae videmus divina et egregia, ipsius scitote esse propria; quae nonnunquam requirimus, ea sunt omnia non a natura, sed a magistro?*[296] Many other axioms and advices there are touching those proprieties and effects, which studies do infuse and instil into manners. And so likewise is there touching the use of all those other points, of company, fame, laws, and the rest, which we recited in the beginning in the doctrine of morality.

14.   But there is a kind of culture of the mind that seemeth yet more accurate and elaborate than the rest, and is built upon this ground; that the minds of all men are at some times in a state more perfect, and at other times in a state more depraved. The purpose therefore of this practice is to fix and cherish the good hours of the mind, and to obliterate and take forth the evil. The fixing of the good hath been practised by two means, vows or constant resolutions, and observances or exercises; which are not to be regarded so much in themselves, as because they keep the mind in continual obedience. The obliteration of the evil hath been practised by two means, some kind of redemption or expiation of that which is past, and an inception or account *de novo,* for the time to come. But this part seemeth sacred and religious, and justly; for all good moral philosophy, as was said, is but a handmaid to religion.

15.   Wherefore we will conclude with that last point, which is of all

other means the most compendious and summary, and again, the most noble and effectual to the reducing of the mind unto virtue and good estate; which is the electing and propounding unto a man's self good and virtuous ends of his life, such as may be in a reasonable sort within his compass to attain. For if these two things be supposed, that a man set before him honest and good ends, and again, that he be resolute, constant, and true unto them; it will follow that he shall mould himself into all virtue at once. And this indeed is like the work of nature; whereas the other course is like the work of the hand. For as when a carver makes an image, he shapes only that part whereupon he worketh (as if he be upon the face, that part which shall be the body is but a rude stone still, till such time as he comes to it); but, contrariwise, when nature makes a flower or living creature, she formeth rudiments of all the parts at one time: so in obtaining virtue by habit, while a man practiseth temperance, he doth not profit much to fortitude, nor the like: but when he dedicateth and applieth himself to good ends, look, what virtue soever the pursuit and passage towards those ends doth commend unto him, he is invested of a precedent disposition to conform himself thereunto. Which state of mind Aristotle doth excellently express himself that it ought not to be called virtuous, but divine: his words are these: *Immanitati autem consentaneum est opponere eam, quae supra humanitatem est, heroicam sive divinam virtutem:* and a little after, *Nam ut ferae neque vitium neque virtus est, sic neque Dei: sed hic quidem status altius quiddam virtute est, ille aliud quiddam a vitio.*[297] And therefore we may see what celsitude of honour Plinius Secundus attributeth to Trajan in his funeral oration;[298] where he said, *That men needed to make no other prayers to the gods, but that they would continue as good Lords to them as Trajan had been;*[299] as if he had not been only an imitation of divine nature, but a pattern of it. But these be heathen and profane passages, having but a shadow of that divine state of mind, which religion and the holy faith doth conduct men unto, by imprinting upon their souls charity, which is excellently called the bond of perfection, because it comprehendeth and fasteneth all virtues together.[300] And as it is elegantly said by Menander of vain love, which is but a false imitation of divine love, *Amor melior Sophista laevo ad humanam vitam,*[301] that love teacheth a man to carry himself better than the sophist or preceptor; which he calleth left-handed, because, with all his rules and preceptions, he cannot form a man so dexterously, nor with that facility to prize himself and govern himself, as love can do: so certainly, if a man's mind be truly inflamed with charity, it doth work him suddenly into a greater perfection than all the doctrine of morality can do, which is but a sophist in comparison of the other. Nay further, as Xenophon observed truly, that all other affections, though they raise the mind, yet they do it by distorting and uncomeliness of ecstasies or excesses; but only love doth exalt the mind, and nevertheless at the same instant doth settle and compose it;[302] so in all other excellencies, though they advance nature, yet they are subject to excess; only charity admitteth no excess. For so we see, aspiring to be like God in power, the angels transgressed and fell; *Ascendam, et ero similis altissimo:*[303] by aspiring to be like God in knowledge, man transgressed and fell; *Eritis sicut Dii, scientes bonum et malum:*[304] but by aspiring to a similitude of God in goodness or love, neither man nor angel ever transgressed, or shall transgress. For unto that imitation we are called: *Diligite inimicos vestros, benefacite eis qui oderunt vos, et orate pro persequentibus et calumniantibus vos, ut sitis filii Patris vestri qui in Caelis est, qui solem suum oriri facit super bonos et malos, et pluit super justos et injustos.*[305] So in the first platform of the divine nature itself, the heathen religion speaketh thus, *Optimus Maximus:* and the sacred Scriptures thus, *Misericordia ejus super omnia opera ejus.*[306]

16. Wherefore I do conclude this part of moral knowledge, concerning the culture and regimen of the mind; wherein if any man, considering the parts thereof which I have enumerated, do judge that my labour is but to collect into an art of science that which hath been pretermitted by others, as matter of common sense and experience, he judgeth well. But as Philocrates sported with Demosthenes, *You may not marvel, Athenians, that Demosthenes and I do differ; for he drinketh water, and I drink wine;*[307] and like as we read of an ancient parable of *the two gates of sleep,*

> Sunt geminae somni portae: quarum altera fertur
> Cornea, qua veris facilis datur exitus umbris:
> Altera candenti perfecta nitens elephanto,
> Sed falsa ad caelum mittunt insomnia manes:[308]

so if we put on sobriety and attention, we shall find it a sure maxim in knowledge, that the more pleasant liquor of wine is the more vaporous, and the braver gate of ivory sendeth forth the falser dreams.

17.   But we have now concluded that general part of human philosophy, which contemplateth man segregate, and as he consisteth of body and spirit. Wherein we may further note, that there seemeth to be a relation or conformity between the good of the mind and the good of the body. For as we divided the good of the body into health, beauty, strength, and pleasure; so the good of the mind, inquired in rational and moral knowledges, tendeth to this, to make the mind sound, and without perturbation; beautiful, and graced with decency; and strong and agile for all duties of life. These three, as in the body, so in the mind, seldom meet, and commonly sever. For it is easy to observe, that many have strength of wit and courage, but have neither health from perturbations, nor any beauty or decency in their doings; some again have an elegancy and fineness of carriage, which have neither soundness of honesty, nor substance of sufficiency: and some again have honest and reformed minds, that can neither become themselves nor manage business: and sometimes two of them meet, and rarely all three. As for pleasure, we have likewise determined that the mind ought not to be reduced to stupid,[309] but to retain pleasure; confined rather in the subject of it, than in the strength and vigour of it.

XXIII   1.   Civil knowledge is conversant about a subject

which of all others is most immersed in matter, and hardliest reduced to axiom. Nevertheless, as Cato the Censor said, *That the Romans were like sheep, for that a man might better drive a flock of them, than one of them; for in a flock, if you could but get some few to go right, the rest would follow:*[310] so in that respect moral philosophy is more difficile than policy. Again, moral philosophy propoundeth to itself the framing of internal goodness; but civil knowledge requireth only an external goodness; for that as to society sufficeth. And therefore it cometh oft to pass that there be evil times in good governments: for so we find in the holy story, when the kings were good, yet it is added, *Sed adhuc populus non direxerat cor suum ad Dominum Deum patrum suorum.*[311] Again, states, as great engines, move slowly, and are not so soon put out of frame: for as in Egypt the seven good years sustained the seven bad, so governments for a time well grounded, do bear out errors following; but the resolution of particular persons is more suddenly subverted. These respects do somewhat qualify the extreme difficulty of civil knowledge.

2.   This knowledge hath three parts, according to the three summary actions of society; which are conversation, negotiation, and government. For man seeketh in society comfort, use, and protection: and they be three wisdoms of divers natures, which do often sever: wisdom of the behaviour, wisdom of business, and wisdom of state.

3.   The wisdom of conversation ought not to be over much affected, but much less despised; for it hath not only an honour in itself, but an influence also into business and government. The poet saith, *Nec vultu destrue verbo tuo:*[312] a man may destroy the force of his words with his countenance: so may he of his deeds, saith Cicero, recommending to his brother affability and easy access; *Nil interest habere ostium apertum, vultum clausum;*[313] it is nothing won to admit men with an open door, and to receive them with a shut and reserved countenance. So we see Atticus, before the first interview between Caesar and Cicero, the war depending, did seriously advise Cicero touching the composing and ordering of his countenance and gesture.[314] And if the government of the countenance be of such effect, much more is that of the speech, and other carriage appertaining to conversation; the true model whereof seemeth to me well expressed by Livy, though not meant for this purpose: *Ne aut arrogans videar, aut obnoxius; quorum alterum est alienae libertatis obliti, alterum suae:*[315] The sum of behaviour is to retain a man's own dignity, without intruding upon the liberty of others. On the other side, if behaviour and outward carriage be intended too much, first it may pass into affectation, and then *Quid deformius quam scenam in vitam transferre,* to act a man's life? But although it proceed not to that extreme, yet it consumeth time, and employeth the mind too much. And therefore as we use to advise young students from company keeping, by saying, *Amici fures temporis:* so certainly the intending of the discretion of behaviour is a great thief of meditation. Again, such as are accomplished in that form of urbanity please themselves in it, and seldom aspire to higher virtue; whereas those

that have defect in it do seek comeliness by reputation; for where reputation is, almost everything becometh; but where that is not, it must be supplied by *puntos,* and compliments. Again, there is no greater impediment of action than an over-curious observance of decency, and the guide of decency, which is time and season. For as Salomon saith, *Qui respicit ad ventos, non seminat; et qui respicit ad nubes, non metet:*[316] a man must make his opportunity, as oft as find it. To conclude, behaviour seemeth to me as a garment of the mind, and to have the conditions of a garment. For it ought to be made in fashion; it ought not to be too curious; it ought to be shaped so as to set forth any good making of the mind, and hide any deformity; and above all, it ought not to be too strait, or restrained for exercise or motion. But this part of civil knowledge hath been elegantly handled, and therefore I cannot report it for deficient.

4.   The wisdom touching negotiation or business hath not been hitherto collected into writing, to the great derogation of learning, and the professors of learning. For from this root springeth chiefly that note or opinion, which by us is expressed in adage to this effect, that there is no great concurrence between learning and wisdom. For of the three wisdoms which we have set down to pertain to civil life, for wisdom of behaviour it is by learned men for the most part despised, as an inferior to virtue, and an enemy to meditation; for wisdom of government, they acquit themselves well, when they are called to it, but that happeneth to few; but for the wisdom of business, wherein man's life is most conversant, there be no books of it, except some few scattered advertisements, that have no proportion to the magnitude of this subject. For if books were written of this, as the other, I doubt not but learned men with mean experience, would far excel men of long experience without learning, and outshoot them in their own bow.

5.   Neither needeth it at all to be doubted, that this knowledge should be so variable as it falleth not under precept; for it is much less infinite than science of government, which, we see, is laboured and in some part reduced. Of this wisdom, it seemeth some of the ancient Romans in the saddest and wisest times were professors; for Cicero reporteth[317] that it was then in use for senators that had name and opinion for general wise men, as Coruncanius, Curius, Laelius, and many others, to

walk at certain hours in the Place, and to give audience to those that would use their advice; and that the particular citizens would resort unto them, and consult with them of the marriage of a daughter, or of the employing of a son, or of a purchase or bargain, or of an accusation, and every other occasion incident to man's life. So as there is a wisdom of counsel and advice even in private causes, arising out of a universal insight into the affairs of the world; which is used indeed upon particular cases propounded, but is gathered by general observation of causes of like nature. For so we see in the book which Q. Cicero writeth to his brother, *De petitione consulatus* (being the only book of business that I know written by the ancients), although it concerned a particular action set on foot, yet the substance thereof consisteth of many wise and politic axioms, which contain not a temporary, but a perpetual direction in the case of popular elections. But chiefly we may see in those aphorisms which have place among divine writings, composed by Salomon the king (of whom the Scriptures testify that his heart was as the sands of the sea,[318] encompassing the world and all worldly matters), we see, I say, not a few profound and excellent cautions, precepts, positions, extending to much variety of occasions; whereupon we will stay awhile, offering to consideration some number of examples.

6.   *Sed et cunctis sermonibus qui dicuntur ne accommodes aurem tuam, ne forte audias servum tuum maledicentem tibi.*[319] Here is concluded the provident stay of inquiry of that which we would be loth to find: as it was judged great wisdom in Pompeius Magnus that he burned Sertorius' papers unperused.[320]

Vir sapiens, si cum stulto contenderit, sive irascatur, sive rideat, non inveniet requiem.[321] Here is described the great disadvantage which a wise man hath in undertaking a lighter person than himself; which is such an engagement as, whether a man turn the matter to jest, or turn it to heat, or howsoever he change copy, he can no ways quit himself well of it.

Qui delicate a pueritia nutrit servum suum, postea sentiet eum contumacem.[322] Here is signified, that if a man begin too high a pitch in his favours, it doth commonly end in unkindness and unthankfulness.

Vidisti virum velocem in opere suo? coram regibus stabit, nec erit inter ignobiles.[323] Here is observed, that of all virtues for rising to honour, quickness of despatch is the best; for superiors many times love not to have those they employ too deep or too sufficient, but ready and diligent.

Vidi cunctos viventes qui ambulant sub sole, cum adolescente secundo qui consurgit pro eo.[324] Here is expressed that which was noted by Sulla first, and after him by Tiberius; *Plures adorant solem orientem quam occidentem vel meridianum.*[325]

Si spiritus potestatem habentis ascenderit super te, locum tuum ne demiseris; quia curatio faciet cessare peccata maxima.[326] Here caution is given, that upon displeasure, retiring is of all courses the unfittest; for a man leaveth things at worst, and depriveth himself of means to make them better.

Erat civitas parva, et pauci in ea viri: venit contra eam rex magnus, et vallavit eam, instruxitque munitiones per gyrum, et perfecta est obsidio; inventusque est in ea vir pauper et sapiens, et liberavit eam per sapientiam suam; et nullus deinceps recordatus est hominis illius pauperis.[327] Here the corruption of states is set forth, that esteem not virtue or merit longer than they have use of it.

Mollis responsio frangit iram.[328] Here is noted that silence or rough answer exasperateth; but an answer present and temperate pacifieth.

Iter pigrorum quasi sepes spinarum.[329] Here is lively represented how laborious sloth proveth in the end; for when things are deferred till the last instant, and nothing prepared beforehand, every step findeth a brier or an impediment, which catcheth or stoppeth.

Melior est finis orationis quam principium.[330] Here is taxed the vanity of formal speakers, that study more about prefaces and inducements, than upon the conclusions and issues of speech.

Qui cognoscit in judicio faciem, non bene facit; iste et pro buccella panis deseret veritatem.[331] Here is noted, that a judge were better be a briber than a respecter of persons; for a corrupt judge offendeth not so lightly as a facile.

Vir pauper calumnians pauperes similis est imbri vehementi, in quo paratur fames.[332] Here is expressed the extremity of necessitous extortions, figured in the ancient fable of the full and the hungry horseleech.

Fons turbatus pede, et vena corrupta, est justus cadens coram impio.[333] Here is noted, that one judicial and exemplar iniquity in the face of the world, doth trouble the fountains of justice more than many particular injuries passed over by connivance.

Qui subtrahit aliquid a patre et a matre, et dicit hoc non esse peccatum, particeps est homicidii.[334] Here is noted, that whereas men in wronging their best friends use to extenuate their fault, as if they might presume or be bold upon them, it doth contrariwise indeed aggravate their fault, and turneth it from injury to impiety.

Noli esse amicus homini iracundo, nec ambulato cum homine furioso.[335] Here caution is given, that in the election of our friends we do principally avoid those which are impatient, as those that will espouse us to many factions and quarrels.

Qui conturbat domum suam, possidebit ventum.[336] Here is noted, that in domestical separations and breaches men do promise to themselves quieting of their mind and contentment; but still they are deceived of their expectation, and it turneth to wind.

Filius sapiens laetificat patrem: filius vero stultus moestitia est matri suae.[337] Here is distinguished, that fathers have most comfort of the good proof of their sons; but mothers have most discomfort of their ill proof, because women have little discerning of virtue, but of fortune.

Qui celat delictum, quaerit amicitiam; sed qui altero sermone repetit, separat foederatos.[338] Here caution is given, that reconcilement is better managed by an amnesty, and passing over that which is past, than by apologies and excusations.

In omni opere bono erit abundantia; ubi autem verba sunt plurima, ibi frequenter egestas.[339] Here is noted, that words and discourse aboundeth most where there is idleness and want.

Primus in sua causa justus; sed venit altera pars, et inquiret in eum.[340] Here is observed, that in all causes the first tale possesseth much; in sort that the prejudice thereby wrought will be hardly removed, except some abuse or falsity in the information be detected.

Verba bilinguis quasi simplicia, et ipsa perveniunt ad interiora ventris.[341] Here is distinguished, that flattery and insinuation, which seemeth set and artificial, sinketh not far; but that entereth deep which hath show of nature, liberty, and simplicity.

Qui erudit derisorem, ipse sibi injuriam facit; et qui arguit impium, sibi maculam generat.[342] Here caution is given how we tender reprehension to arrogant and scornful natures, whose manner is to esteem it for contumely, and accordingly to return it.

Da sapienti occasionem, et addetur ei sapientia.[343] Here is distinguished the wisdom brought into habit, and that which is but verbal, and swimming only in conceit; for the one upon the occasion presented is quickened and redoubled, the other is amazed and confused.

Quomodo in aquis resplendent vultus prospicientium, sic corda hominum manifesta sunt prudentibus.[344] Here the mind of a wise man is compared to a glass, wherein the images of all diversity of natures and customs are represented; from which representation proceedeth that application,

Qui sapit, innumeris moribus aptus erit.[345]

7.   Thus have I stayed somewhat longer upon these sentences politic of Salomon than is agreeable to the proportion of an example; led with a desire to give authority to this part of knowledge, which I noted as deficient, by so excellent a precedent; and have also attended them with brief observations, such as to my understanding offer no violence to the sense, though I know they may be applied to a more divine use: but it is allowed, even in divinity, that some interpretations, yea, and some writings, have more of the eagle than others;[346] but taking them as instructions for life, they might have received large discourse, if I would have broken them and illustrated them by deducements and examples.

8.   Neither was this in use only with the Hebrews, but it is generally to be found in the wisdom of the more ancient times; that as men found out any observation that they thought was good for life, they would gather it, and express it in parable, or aphorism, or fable. But for fables, they were vicegerents and supplies where examples failed: now that the times abound with history, the aim is better when the mark is alive. And therefore the form of writing which of all others is fittest for this variable argument of negotiation and occasions is that which Machiavel chose wisely and aptly for government; namely, discourse upon histories or examples. For knowledge drawn freshly, and in our view, out of particulars, knoweth the way best to particulars again; and it hath much greater life for practice when the discourse attendeth upon the example, than when the example attendeth upon

the discourse. For this is no point of order, as it seemeth at first, but of substance: for when the example is the ground, being set down in a history at large, it is set down with all circumstances, which may sometimes control the discourse thereupon made, and sometimes supply it as a very pattern for action; whereas the examples alleged for the discourse's sake are cited succinctly, and without particularity, and carry a servile aspect towards the discourse which they are brought in to make good.

9.   But this difference is not amiss to be remembered, that as history of Times is the best ground for discourse of government, such as Machiavel handleth, so history of Lives, because it is the most proper for discourse of business, is more conversant in private actions. Nay, there is a ground of discourse for this purpose fitter than them both, which is discourse upon letters, such as are wise and weighty, as many are of Cicero *ad Atticum,* and others. For letters have a great and more particular representation of business than either chronicles or lives. Thus have we spoken both of the matter and form of this part of civil knowledge, touching negotiation, which we note to be deficient.

10.   But yet there is another part of this part, which differeth as much from that whereof we have spoken as *sapere* and *sibi sapere,* the one moving as it were to the circumference, the other to the centre. For there is a wisdom of counsel, and again there is a wisdom of pressing a man's own fortune; and they do sometimes meet, and often sever. For many are wise in their own ways that are weak for government or counsel; like ants, which is a wise creature for itself, but very hurtful for the garden. This wisdom the Romans did take much knowledge of: *Nam pol sapiens,* saith the comical poet, *fingit fortunam sibi;*[347] and it grew to an adage, *Faber quisque fortunae propriae;*[348] and Livy attributed it to Cato the first, *in hoc viro tanta vis animi et ingenii inerat, ut quocunque loco natus esset sibi ipse fortunam facturus videretur.*[349]

11.   This conceit or position, if it be too much declared and professed, hath been thought a thing impolitic and unlucky, as was observed in Timotheus the Athenian, who, having done many great services to the estate in his government, and giving an account thereof to the people, as the manner was, did conclude every particular with this clause, *And in this fortune had no part.*[350] And it came so to pass, that he never pros-

pered in any thing he took in hand afterwards: for this is too high and too arrogant, savouring of that which Ezekiel saith of Pharaoh, *Dicis, Fluvius est meus et ego feci memet ipsum:*[351] or of that which another prophet speaketh, that men offer sacrifices to their nets and snares:[352] and that which the poet expresseth,

> Dextra mihi Deus, et telum quod missile libro,
> Nunc adsint![353]

for these confidences were ever unhallowed, and unblessed: and therefore those that were great politiques indeed ever ascribed their successes to their felicity, and not to their skill or virtue. For so Sulla surnamed himself Felix, not Magnus: so Caesar said to the master of the ship, *Caesarem portas et fortunam ejus.*[354]

12. But yet nevertheless these positions, *Faber quisque fortunae suae: sapiens dominabitur astris:*[355] *invia virtuti nulla est via,*[356] and the like, being taken and used as spurs to industry, and not as stirrups to insolency, rather for resolution than for presumption or outward declaration, have been ever thought sound and good; and are, no question, imprinted in the greatest minds, who are so sensible of this opinion, as they can scarce contain it within. As we see in Augustus Caesar (who was rather diverse from his uncle, than inferior in virtue), how when he died, he desired his friends about him to give him a *plaudite,* as if he were conscient to himself that he had played his part well upon the stage.[357] This part of knowledge we do report also as deficient: not but that it is practised too much, but it hath not been reduced to writing. And therefore lest it should seem to any that it is not comprehensible by axiom, it is requisite, as we did in the former, that we set down some heads or passages of it.

13. Wherein it may appear at the first a new and unwonted argument to teach men how to raise and make their fortune; a doctrine wherein every man perchance will be ready to yield himself a disciple, till he see the difficulty; for fortune layeth as heavy impositions as virtue; and it is as hard and severe a thing to be a true politique, as to be truly moral. But the handling hereof concerneth learning greatly, both in honour and in substance: in honour, because pragmatical men may not go away with an opinion that learning is like a lark, that can mount, and sing, and please herself, and nothing else; but may know that she holdeth as well of the hawk, that can soar aloft, and can also descend and strike upon the prey: in substance, because it is the perfect law of inquiry of truth, that nothing be in the globe of matter, which should not be likewise in the globe of crystal, or form; that is, that there be not any thing in being and action, which should not be drawn and collected into contemplation and doctrine. Neither doth learning admire or esteem of this architecture of fortune, otherwise than as of an inferior work: for no man's fortune can be an end worthy of his being; and many times the worthiest men do abandon their fortune willingly for better respects: but nevertheless fortune, as an organ of virtue and merit, deserveth the consideration.

14. First, therefore, the precept which I conceive to be most summary towards the prevailing in fortune, is to obtain that window which Momus did require:[358] who seeing in the frame of man's heart such angles and recesses, found fault that there was not a window to look into them; that is, to procure good informations of particulars touching persons, their natures, their desires and ends, their customs and fashions, their helps and advantages, and whereby they chiefly stand: so again their weaknesses and disadvantages, and where they lie most open and obnoxious; their friends, factions, dependencies; and again their opposites, enviers, competitors, their moods and times, *Sola viri molles aditus et tempora noras;*[359] their principles, rules, and observations, and the like: and this not only of persons, but of actions; what are on foot from time to time, and how they are conducted, favoured, opposed, and how they import, and the like. For the knowledge of present actions is not only material in itself, but without it also the knowledge of persons is very erroneous: for men change with the actions; and whiles they are in pursuit they are one, and when they return to their nature they are another. These informations of particulars, touching persons and actions, are as the minor propositions in every active syllogism; for no excellency of observations, which are as the major propositions, can suffice to ground a conclusion, if there be error and mistaking in the minors.

15. That this knowledge is possible, Salomon is our surety; who saith, *Consilium in corde viri tanquam aqua profunda; sed vir prudens exhauriet illud.*[360] And although the knowledge itself falleth not under precept, because it is of individuals, yet the instructions for the obtaining of it may.

16. We will begin, therefore, with this precept, according to the ancient opinion, that the sinews of wisdom are slowness of belief and distrust; that more trust be given to countenances and deeds than to words: and in words rather to sudden passages and surprised words than to set and purposed words. Neither let that be feared which is said, *Fronti nulla fides:*[361] which is meant of a general outward behaviour, and not of the private and subtle motions and labours of the countenance and gesture; which as Q. Cicero elegantly saith, is *Animi janua, the gate of the mind.*[362] None more close than Tiberius, and yet Tacitus saith of Gallus, *Etenim vultu offensionem conjectaverat.*[363] So again, noting the differing character and manner of his commending Germanicus and Drusus in the senate, he saith, touching his fashion wherein he carried his speech of Germanicus, thus; *Magis in speciem adornatis verbis, quam ut penitus sentire crederetur:* but of Drusus thus: *Paucioribus, sed intentior, et fida oratione:*[364] and in another place, speaking of his character of speech, when he did any thing that was gracious and popular, he saith, that in other things he was *velut eluctantium verborum;* but then again, *solutius loquebatur quando subveniret.*[365] So that there is no such artificer of dissimulation, nor no such commanded countenance, *vultus jussus,* that can sever from a feigned tale some of these fashions, either a more slight and careless fashion, or more set and formal, or more tedious and wandering, or coming from a man more drily and hardly.

17. Neither are deeds such assured pledges, as that they may be trusted without a judicious consideration of their magnitude and nature: *Fraus sibi in parvis fidem praestruit, ut majore emolumento fallat:*[366] and the Italian thinketh himself upon the point to be bought and sold, when he is better used than he was wont to be, without manifest cause. For small favours, they do but lull men asleep, both as to caution and as to industry; and are, as Demosthenes calleth them, *Alimenta socordiae.*[367] So again we see how false the nature of some deeds are, in that particular which Mutianus practised upon Antonius Primus, upon that hollow and unfaithful reconcilement which was made between them; whereupon Mutianus advanced many of the friends of

Antonius: *simul amicis ejus praefecturas et tribunatus largitur:*[368] wherein, under pretence to strengthen him, he did desolate him and won from him his dependences.

18.   As for words, though they be like waters to physicians, full of flattery and uncertainty, yet they are not to be despised, especially with the advantage of passion and affection. For so we see Tiberius, upon a stinging and incensing speech of Agrippina, came a step forth of his dissimulation, when he said, *You are hurt because you do not reign;* of which Tacitus saith, *Audita haec raram occulti pectoris vocem elicuere; correptamque Graeco versu admonuit, ideo laedi, quia non regnaret.*[369] And therefore the poet doth elegantly call passions tortures, that urge men to confess their secrets:

> Vino tortus et ira.[370]

And experience showeth, there are few men so true to themselves and so settled, but that, sometimes upon heat, sometimes upon bravery, sometimes upon kindness, sometimes upon trouble of mind and weakness, they open themselves; especially if they be put to it with a counter-dissimulation, according to the proverb of Spain, *Di mentira, y sacaras verdad* (*Tell a lie and find a truth*).

19.   As for the knowing of men which is at second hand from reports; men's weaknesses and faults are best known from their enemies, their virtues and abilities from their friends, their customs and times from their servants, their conceits and opinions from their familiar friends, with whom they discourse most. General fame is light, and the opinions conceived by superiors or equals are deceitful; for to such men are more masked: *Verior fama e domesticis emanat.*[371]

20.   But the soundest disclosing and expounding of men is by their natures and ends, wherein the weakest sort of men are best interpreted by their natures, and the wisest by their ends. For it was both pleasantly and wisely said, though I think very untruly, by a nuncio of the pope, returning from a certain nation where he served as lidger; whose opinion being asked touching the appointment of one to go in his place, he wished that in any case they did not send one that was too wise; because no very wise man would ever imagine what they in that country were like to do. And certainly it is an error frequent for men to shoot over, and to suppose deeper ends and more compass-reaches than are: the Italian proverb being elegant, and for the most part true:

Di danari, di senno, e di fede,
C'è ne manco che non credi.

There is commonly less money, less wisdom, and less good faith than men do account upon.

21.   But princes, upon a far other reason, are best interpreted by their natures, and private persons by their ends. For princes being at the top of human desires, they have for the most part no particular ends whereto they aspire, by distance from which a man might take measure and scale of the rest of their actions and desires; which is one of the causes that maketh their hearts more inscrutable.[372] Neither is it sufficient to inform ourselves in men's ends and natures, of the variety of them only, but also of the predominancy, what humour reigneth most, and what end is principally sought. For so we see, when Tigellinus saw himself outstripped by Petronius Turpilianus in Nero's humours of pleasures, *metus ejus rimatur,*[373] he wrought upon Nero's fears, whereby he brake the other's neck.

22.   But to all this part of inquiry the most compendious way resteth in three things: the first, to have general acquaintance and inwardness with those which have general acquaintance and look most into the world; and especially according to the diversity of business, and the diversity of persons, to have privacy and conversation with some one friend at least which is perfect and well intelligenced in every several kind. The second is, to keep a good mediocrity in liberty of speech and secrecy; in most things liberty; secrecy where it importeth; for liberty of speech inviteth and provoketh liberty to be used again, and so bringeth much to a man's knowledge; and secrecy, on the other side, induceth trust and inwardness. The last is, the reducing of a man's self to this watchful and serene habit, as to make account and purpose, in every conference and action, as well to observe as to act. For as Epictetus would have a philosopher in every particular action to say to himself, *Et hoc volo, et etiam institutum servare,*[374] so a politic man in everything should say to himself, *Et hoc volo, ac etiam aliquid addiscere.* I have stayed the longer upon this precept of obtaining good information,

because it is a main part by itself, which answereth to all the rest. But, above all things, caution must be taken that men have a good stay and hold of themselves, and that this much knowing do not draw on much meddling; for nothing is more unfortunate than light and rash inter-meddling in many matters. So that this variety of knowledge tendeth in conclusion but only to this, to make a better and freer choice of those actions which may concern us, and to conduct them with the less error and the more dexterity.

23.   The second precept concerning this knowledge is, for men to take good information touching their own person, and well to under-stand themselves: knowing that, as St. James saith, though men look oft in a glass,[375] yet they do suddenly forget themselves; wherein as the divine glass is the word of God, so the politic glass is the state of the world, or times wherein we live, in the which we are to behold ourselves.

24.   For men ought to take an unpartial view of their own abilities and virtues; and again of their wants and impediments; accounting these with the most, and those other with the least; and from this view and examination to frame the considerations following.

25.   First, to consider how the constitution of their nature sorteth with the general state of the times; which if they find agreeable and fit, then in all things to give themselves more scope and liberty; but if differing and dissonant, then in the whole course of their life to be more close, retired, and reserved: as we see in Tiberius, who was never seen at a play, and came not into the Senate in twelve of his last years; whereas Augustus Caesar lived ever in men's eyes, which Tacitus observ-eth, *alia Tiberio morum via.*[376]

26.   Secondly, to consider how their nature sorteth with professions and courses of life, and accordingly to make election, if they be free; and, if engaged, to make the departure at the first opportunity: as we see was done by Duke Valentine,[377] that was designed by his father to a sacerdotal profession, but quitted it soon after in regard of his parts and inclination; being such, nevertheless, as a man cannot tell well whether they were worse for a prince or for a priest.

27.   Thirdly, to consider how they sort with those whom they are like to have competitors and concurrents; and to take that course wherein there is most solitude, and themselves like to be most eminent:

as Caesar Julius did, who at first was an orator or pleader; but when he saw the excellency of Cicero, Hortensius, Catulus, and others, for eloquence, and saw there was no man of reputation for the wars but Pompeius, upon whom the state was forced to rely, he forsook his course begun towards a civil and popular greatness and transferred his designs to a martial greatness.

28.　Fourthly, in the choice of their friends and dependences, to proceed according to the composition of their own nature: as we may see in Caesar; all whose friends and followers were men active and effectual, but not solemn, or of reputation.

29.　Fifthly, to take special heed how they guide themselves by examples, in thinking they can do as they see others do; whereas perhaps their natures and carriages are far differing. In which error it seemeth Pompey was, of whom Cicero saith, that he was wont often to say, *Sulla potuit, ego non potero?*[378] Wherein he was much abused, the natures and proceedings of himself and his example being the unlikest in the world; the one being fierce, violent, and pressing the fact; the other solemn, and full of majesty and circumstance, and therefore the less effectual.

But this precept touching the politic knowledge of ourselves, hath many other branches, whereupon we cannot insist.

30.　Next to the well understanding and discerning of a man's self, there followeth the well opening and revealing a man's self; wherein we see nothing more usual than for the more able man to make the less show. For there is a great advantage in the well setting forth of a man's virtues, fortunes, merits; and again, in the artificial covering of a man's weaknesses, defects, disgraces; staying upon the one, sliding from the other; cherishing the one by circumstances, gracing the other by exposition, and the like: wherein we see what Tacitus saith of Mutianus, who was the greatest politique of his time, *Omnium quae dixerat feceratque arte quadam ostentator:*[379] which requireth indeed some art, lest it turn tedious and arrogant; but yet so as ostentation, though it be to the first degree of vanity, seemeth to me rather a vice in manners than in policy: for as it is said, *Audacter calumniare, semper aliquid haeret:*[380] so, except it be in a ridiculous degree of deformity, *Audacter te vendita, semper aliquid haeret.* For it will stick with the more ignorant and inferior sort of men, though men of wisdom and rank do smile at it, and despise it; and yet the authority won with many doth countervail the disdain of a few. But if it be carried with decency and government, as with a natural, pleasant, and ingenious fashion; or at times when it is mixed with some peril and unsafety, as in military persons; or at times when others are most envied; or with easy and careless passage to it and from it, without dwelling too long, or being too serious; or with an equal freedom of taxing a man's self, as well as gracing himself; or by occasion of repelling or putting down others' injury or insolence; it doth greatly add to reputation: and surely not a few solid natures, that want this ventosity, and cannot fail in the height of the winds, are not without some prejudice and disadvantage by their moderation.

31.   But for these flourishes and enhancements of virtue, as they are not perchance unnecessary, so it is at least necessary that virtue be not disvalued and imbased under the just price; which is done in three manners: by offering and obtruding a man's self; wherein men think he is rewarded when he is accepted; by doing too much, which will not give that which is well done leave to settle, and in the end induceth satiety; and by finding too soon the fruit of a man's virtue, in commendation, applause, honour, favour; wherein if a man be pleased with a little, let him hear what is truly said: *Cave ne insuetus rebus majoribus videaris, si haec te res parva sicuti magna delectat* [381]

32.   But the covering of defects is of no less importance than the valuing of good parts; which may be done likewise in three manners, by *caution,* by *colour,* and by *confidence. Caution* is when men do ingeniously and discreetly avoid to be put into those things for which they are not proper: whereas, contrariwise, bold and unquiet spirits will thrust themselves into matters without difference, and so publish and proclaim all their wants. *Colour* is, when men make a way for themselves, to have a construction made of their faults or wants, as proceeding from a better cause, or intended for some other purpose: for of the one it is well said,

Saepe latet vitium proximitate boni,[382]

and therefore whatsoever want a man hath, he must see that he pretend the virtue that shadoweth it; as if he be dull, he must affect gravity; if a coward, mildness; and so the rest: for the second, a man must frame some probable cause why he should not do his best, and why he should dissemble his abilities; and for that purpose must use to dissemble those abilities which are notorious in him, to give colour that his true wants are but industries and dissimulations. For *confidence,* it is the last[383] but surest remedy; namely, to depress and seem to despise whatsoever a man cannot attain; observing the good principle of the merchants, who endeavour to raise the price of their own commodities, and to beat down the price of others. But there is a confidence that passeth this other; which is, to face out a man's own defects, in seeming to conceive that he is best in those things wherein he is failing; and, to help that again, to seem on the other side that he hath least opinion of himself in those things wherein he is best: like as we shall see it commonly in poets, that if they show their verses, and you except to any, they will say, *that that line cost them more labour than any of the rest;* and presently will seem to disable and suspect rather some other line, which they know well enough to be the best in the number. But above all, in this righting and helping of a man's self in his own carriage, he must take heed he show not himself dismantled, and exposed to scorn and injury, by too much dulceness, goodness, and facility of nature; but show some sparkles of liberty, spirit, and edge. Which kind of fortified carriage, with a ready rescuing of a man's self from scorns, is sometimes of necessity imposed upon men by somewhat in their person or fortune; but it ever succeedeth with good felicity.

33.   Another precept of this knowledge is, by all possible endeavour to frame the mind to be pliant and obedient to occasion; for nothing hindereth men's fortunes so much as this: *Idem manebat, neque idem decebat,*[384] men are where they were, when occasions turn: and therefore to Cato, whom Livy maketh such an architect of fortune, he addeth, that he had *versatile ingenium.*[385] And thereof it cometh that these grave solemn wits, which must be like themselves, and cannot make departures, have more dignity than felicity. But in some it is nature to be somewhat viscous and inwrapped, and not easy to turn; in some it is a conceit, that is almost a nature, which is, that men can hardly make themselves believe that they ought to change their course, when they have found good by it in former experience. For Machiavel noted wisely,

how Fabius Maximus would have been temporizing still, according to his old bias, when the nature of the war was altered and required hot pursuit.[386] In some other it is want of point and penetration in their judgment, that they do not discern when things have a period, but come in too late after the occasion; as Demosthenes[387] compareth the people of Athens to country fellows, when they play in a fence school, that if they have a blow, then they remove their weapon to that ward, and not before. In some other it is a lothness to leese labours passed, and a conceit that they can bring about occasions to their ply; and yet in the end, when they see no other remedy, then they come to it with disadvantage; as Tarquinius, that gave for the third part of Sibylla's books the treble price,[388] when he might at first have had all three for the simple. But from whatsoever root or cause this restiveness of mind proceedeth, it is a thing most prejudicial; and nothing is more politic than to make the wheels of our mind concentric and more voluble with the wheels of fortune.

34.   Another precept of this knowledge, which hath some affinity with that we last spake of, but with difference, is that which is well expressed, *Fatis accede Deisque*,[389] that men do not only turn with the occasions, but also run with the occasions, and not strain their credit or strength to over-hard or extreme points; but choose in their actions that which is most passable: for this will preserve men from foil, not occupy them too much about one matter, win opinion of moderation, please the most, and make a show of a perpetual felicity in all they undertake; which cannot but mightily increase reputation.

35.   Another part of this knowledge seemeth to have some repugnancy with the former two, but not as I understand it; and it is that which Demosthenes uttereth in high terms; *Et quemadmodum receptum est, ut exercitum ducat imperator, sic et a cordatis viris res ipsae ducendae; ut quae ipsis videntur, ea gerantur, et non ipsi eventus tantum persequi cogantur.*[390] For, if we observe, we shall find two differing kinds of sufficiency in managing of business; some can make use of occasions aptly and dexterously, but plot little; some can urge and pursue their own plots well, but cannot accommodate nor take in;[391] either of which is very unperfect without the other.

36.   Another part of this knowledge is the observing a good medioc-

rity in the declaring, or not declaring a man's self: for although depth of secrecy, and making ways, *qualis est via navis in mari*[392] (which the French calleth *sourdes menées,* when men set things in work without opening themselves at all), be sometimes both prosperous and admirable; yet many times *dissimulatio errores parit, qui dissimulatorem ipsum illaqueant;* and therefore, we see the greatest politiques have in a natural and free manner professed their desires, rather than being reserved and disguised in them. For so we see that Lucius Sulla made a kind of profession, *that he wished all men happy or unhappy, as they stood his friends or enemies.* So Caesar, when he went first into Gaul, made no scruple to profess *that he had rather be first in a village, than second at Rome.*[393] So again, as soon as he had begun the war, we see what Cicero saith of him, *Alter* (meaning of Caesar) *non recusat, sed quodammodo postulat, ut (ut est) sic appelletur tyrannus.*[394] So we may see in a letter of Cicero to Atticus, that Augustus Caesar, in his very entrance into affairs, when he was a darling of the senate, yet in his harangues to the people would swear, *Ita parentis honores consequi liceat,*[395] which was no less than the tyranny; save that, to help it, he would stretch forth his hand towards a statue of Caesar's that was erected in the place: and men laughed, and wondered, and said, Is it possible? or, Did you ever hear the like? and yet thought he meant no hurt; he did it so handsomely and ingenuously. And all these were prosperous: whereas Pompey, who tended to the same ends, but in a more dark and dissembling manner, as Tacitus saith of him, *Occultior, non melior,*[396] wherein Sallust concurreth, *ore probo, animo inverecundo,*[397] made it his design, by infinite secret engines, to cast the state into an absolute anarchy and confusion, that the state might cast itself into his arms for necessity and protection, and so the sovereign power be put upon him, and he never seen in it: and when he had brought it, as he thought, to that point, when he was chosen consul alone, as never any was, yet he could make no great matter of it, because men understood him not; but was fain, in the end, to go the beaten track of getting arms into his hands, by colour of the doubt of Caesar's designs: so tedious, casual, and unfortunate are these deep dissimulations: whereof it seemeth Tacitus made his judgment, that they were a cunning of an inferior form in regard of true policy; attributing the one to Augustus, the other to Tiberius; where speaking of Livia, he saith, *Et cum artibus mariti simulatione filii bene composita:*[398] for surely the continual habit of dissimulation is but a weak and sluggish cunning, and not greatly politic.

37. Another precept of this architecture of fortune is, to accustom our minds to judge of the proportion or value of things, as they conduce and are material to our particular ends: and that to do substantially, and not superficially. For we shall find the logical part, as I may term it, of some men's minds good, but the mathematical part erroneous; that is, they can well judge of consequences, but not of proportions and comparison, preferring things of show and sense before things of substance and effect. So some fall in love with access to princes, others with popular fame and applause, supposing they are things of great purchase: when in many cases they are but matters of envy, peril, and impediment. So some measure things according to the labour and difficulty, or assiduity, which are spent about them; and think, if they be ever moving, that they must needs advance and proceed; as Caesar saith in a despising manner of Cato the second, when he describeth how laborious and indefatigable he was to no great purpose; *Haec omnia magno studio agebat.*[399] So in most things men are ready to abuse themselves in thinking the greatest means to be best, when it should be the fittest.

38. As for the true marshalling of men's pursuits towards their fortune, as they are more or less material, I hold them to stand thus: first the amendment of their own minds. For the remove of the impediments of the mind will sooner clear the passages of fortune, than the obtaining fortune will remove the impediments of the mind. In the second place, I set down wealth and means; which I know most men would have placed first, because of the general use which it beareth towards all variety of occasions. But that opinion I may condemn with like reason as Machiavel[400] doth that other, that moneys were the sinews of the wars; whereas, saith he, the true sinews of the wars are the sinews of men's arms, that is, a valiant, populous, and military nation: and he voucheth aptly the authority of Solon, who, when Croesus showed him his treasury of gold, said to him, that if another came that had better iron, he would be master of his gold. In like manner it may be truly affirmed, that it is not moneys that are the sinews of fortune, but

it is the sinews and steel of men's minds, wit, courage, audacity, resolution, temper, industry, and the like. In the third place I set down reputation, because of the peremptory tides and currents it hath; which, if they be not taken in their due time, are seldom recovered, it being extreme hard to play an after game of reputation. And lastly, I place honour, which is more easily won by any of the other three, much more by all, than any of them can be purchased by honour. To conclude this precept, as there is order and priority in matter, so is there in time, the preposterous placing whereof is one of the commonest errors: while men fly to their ends when they should intend their beginnings, and do not take things in order of time as they come on, but marshal them according to greatness, and not according to instance; not observing the good precept, *Quod nunc instat agamus.*[401]

39. Another precept of this knowledge is not to embrace any matters which do occupy too great a quantity of time, but to have that sounding in a man's ears, *Sed fugit interea, fugit irreparabile tempus:*[402] and that is the cause why those which take their course of rising by professions of burden, as lawyers, orators, painful divines, and the like, are not commonly so politic for their own fortune, otherwise than in their ordinary way, because they want time to learn particulars, to wait occasions, and to devise plots.

40. Another precept of this knowledge is, to imitate nature, which doth nothing in vain; which surely a man may do if he do well interlace his business, and bend not his mind too much upon that which he principally intendeth. For a man ought in every particular action so to carry the motions of his mind, and so to have one thing under another, as if he cannot have that he seeketh in the best degree, yet to have it in a second, or so in a third; and if he can have no part of that which he purposed, yet to turn the use of it to somewhat else; and if he cannot make anything of it for the present, yet to make it as a seed of somewhat in time to come; and if he can contrive no effect or substance from it, yet to win some good opinion by it, or the like. So that he should exact an account of himself of every action, to reap somewhat, and not to stand amazed and confused if he fail of that he chiefly meant: for nothing is more impolitic than to mind actions wholly one by one. For he that doth so leeseth infinite occasions which intervene, and are

many times more proper and propitious for somewhat that he shall need afterwards, than for that which he urgeth for the present; and therefore men must be perfect in that rule, *Haec oportet facere, et illa non omittere.*[403]

41. Another precept of this knowledge is, not to engage a man's self peremptorily in any thing, though it seem not liable to accident; but ever to have a window to fly out at, or a way to retire: following the wisdom in the ancient fable of the two frogs, which consulted when their plash was dry whither they should go; and the one moved to go down into a pit, because it was not likely the water would dry there; but the other answered, *True, but if it do, how shall we get out again?*

42. Another precept of this knowledge is, that ancient precept of Bias, construed not to any point of perfidiousness, but only to caution and moderation, *Et ama tanquam inimicus futurus, et odi tanquam amaturus;*[404] for it utterly betrayeth all utility for men to embark themselves too far in unfortunate friendships, troublesome spleens, and childish and humorous envies or emulations.

43. But I continue this beyond the measure of an example; led, because I would not have such knowledges, which I note as deficient, to be thought things imaginative or in the air, or an observation or two much made of, but things of bulk and mass, whereof an end is hardlier made than a beginning. It must be likewise conceived, that in these points which I mention and set down, they are far from complete tractates of them, but only as small pieces for patterns. And lastly, no man, I suppose, will think that I mean fortunes are not obtained without all this ado; for I know they come tumbling into some men's laps; and a number obtain good fortunes by diligence in a plain way, little intermeddling, and keeping themselves from gross errors.

44. But as Cicero, when he setteth down an idea of a perfect orator, doth not mean that every pleader should be such; and so likewise, when a prince or a courtier hath been described by such as have handled those subjects, the mould hath used to be made according to the perfection of the art, and not according to common practice: so I understand it, that it ought to be done in the description of a politic man, I mean politic for his own fortune.

45. But it must be remembered all this while, that the precepts which

we have set down are of that kind which may be counted and called *Bonae Artes.* As for evil arts, if a man would set down for himself that principle of Machiavel,[405] *that a man seek not to attain virtue itself, but the appearance only thereof; because the credit of virtue is a help, but the use of it is cumber:* or that other of his principles, *that he presuppose, that men are not fitly to be wrought otherwise but by fear; and therefore that he seek to have every man obnoxious, low, and in strait,* which the Italians call *seminar spine,* to sow thorns: or that other principle, contained in the verse which Cicero citeth, *Cadant amici, dummodo inimici intercidant,*[406] as the triumvirs, which sold, every one to other, the lives of their friends for the deaths of their enemies: or that other protestation of L. Catilina, to set on fire and trouble states, to the end to fish in droumy waters, and to unwrap their fortunes, *Ego si quid in fortunis meis excitatum sit incendium, id non aqua sed ruina restinguam:*[407] or that other principle of Lysander, *that children are to be deceived with comfits, and men with oaths:*[408] and the like evil and corrupt positions, whereof, as in all things, there are more in number than of the good: certainly with these dispensations from the laws of charity and integrity, the pressing of a man's fortune may be more hasty and compendious. But it is in life as it is in ways, the shortest way is commonly the foulest, and surely the fairer way is not much about.

46.   But men, if they be in their own power, and do bear and sustain themselves, and be not carried away with a whirlwind or tempest of ambition, ought, in the pursuit of their own fortune, to set before their eyes not only that general map of the world, *that all things are vanity and vexation of spirit,*[409] but many other more particular cards and directions: chiefly that, that being without well-being is a curse, and the greater being the greater curse; and that all virtue is most rewarded, and all wickedness most punished in itself: according as the poet saith excellently:

> Quae vobis, quae digna, viri, pro laudibus istis
> Praemia posse rear solvi? pulcherrima primum
> Dii *moresque* dabunt vestri.[410]

And so of the contrary. And, secondly, they ought to look up to the eternal providence and divine judgment, which often subverteth the wisdom of evil plots and imaginations, according to that Scripture, *He hath conceived mischief, and shall bring forth a vain thing.*[411] And although men should refrain themselves from injury and evil arts, yet this incessant and Sabbathless pursuit of a man's fortune leaveth not the tribute which we owe to God of our time; who we see demandeth a tenth of our substance, and a seventh, which is more strict, of our time: and it is to small purpose to have an erected face towards heaven, and a perpetual grovelling spirit upon earth, eating dust, as doth the serpent, *Atque affigit humo divinae particulam aurae.*[412] And if any man flatter himself that he will employ his fortune well, though he should obtain it ill, as was said concerning Augustus Caesar, and after of Septimius Severus, *that either they should never have been born, or else they should never have died,*[413] they did so much mischief in the pursuit and ascent of their greatness, and so much good when they were established; yet these compensations and satisfactions are good to be used, but never good to be purposed. And lastly, it is not amiss for men in their race toward their fortune, to cool themselves a little with that conceit which is elegantly expressed by the Emperor Charles the Fifth, in his instructions to the king his son, *That fortune hath somewhat of the nature of a woman, that if she be too much wooed, she is the farther off.*[414] But this last is but a remedy for those whose tastes are corrupted: let men rather build upon that foundation which is a cornerstone of divinity and philosophy, wherein they join close, namely that same *Primum quaerite.* For divinity saith, *Primum quaerite regnum Dei, et ista omnia adjicientur vobis:*[415] and philosophy saith, *Primum quaerite bona animi; caetera aut aderunt, aut non oberunt.* And although the human foundation hath somewhat of the sands, as we see in M. Brutus, when he brake forth into that speech,

Te colui, Virtus, ut rem; at tu nomen inane es;[416]

yet the divine foundation is upon the rock. But this may serve for a taste of that knowledge which I noted as deficient.

47.   Concerning Government,[417] it is a part of knowledge secret and retired, in both these respects in which things are deemed secret; for some things are secret because they are hard to know, and some because they are not fit to utter. We see all governments are obscure and invisible:

> Totamque infusa per artus
> Mens agitat molem, et magno se corpore miscet.[418]

Such is the description of governments. We see the government of God over the world is hidden, inasmuch as it seemeth to participate of much irregularity and confusion: the government of the soul in moving the body is inward and profound, and the passages thereof hardly to be reduced to demonstration. Again, the wisdom of antiquity (the shadows whereof are in the poets), in the description of torments and pains, next unto the crime of rebellion, which was the giants' offence, doth detest the offence of futility, as in Sisyphus and Tantalus.[419] But this was meant of particulars: nevertheless even unto the general rules of discourses of policy and government there is due a reverent and reserved handling.

48. But contrariwise, in the governors toward the governed, all things ought, as far as the frailty of man permitteth, to be manifest and revealed. For so it is expressed in the Scriptures touching the government of God, that this globe, which seemeth to us a dark and shady body, is in the view of God as crystal: *Et in conspectu sedis tanquam mare vitreum simile crystallo.*[420] So unto princes and states, especially towards wise senates and councils, the natures and dispositions of the people, their conditions and necessities, their factions and combinations, their animosities and discontents, ought to be, in regard of the variety of their intelligences, the wisdom of their observations, and the height of their station where they keep sentinel, in great part clear and transparent. Wherefore, considering that I write to a King that is a master of this science, and is so well assisted, I think it decent to pass over this part in silence, as willing to obtain the certificate which one of the ancient philosophers aspired unto; who being silent, when others contended to make demonstration of their abilities by speech, desired it might be certified for his part, *that there was one that knew how to hold his peace.*

49. Notwithstanding, for the more public part of government, which is laws, I think good to note only one deficiency; which is, that all those which have written of laws, have written either as philosophers or as lawyers, and none as statesmen. As for the philosophers, they make imaginary laws for imaginary commonwealths; and their discourses are as the stars, which give little light, because they are so high. For the lawyers, they write according to the states where they live, what is received law, and not what ought to be law: for the wisdom of a lawmaker is one, and of a lawyer is another. For there are in nature certain fountains of justice, whence all civil laws are derived but as streams: and like as waters do take tinctures and tastes from the soils through which they run, so do civil laws vary according to the regions and governments where they are planted, though they proceed from the same fountains. Again, the wisdom of a lawmaker consisteth not only in a platform of justice, but in the application thereof; taking into consideration by what means laws may be made certain, and what are the causes and remedies of the doubtfulness and incertainty of law; by what means laws may be made apt and easy to be executed, and what are the impediments and remedies in the execution of laws; what influence laws touching private right of *meum* and *tuum* have into the public state, and how they may be made apt and agreeable; how laws are to be penned and delivered, whether in *texts* or in *acts,* brief or large, with preambles, or without; how they are to be pruned and reformed from time to time, and what is the best means to keep them from being too vast in volumes, or too full of multiplicity and crossness; how they are to be expounded, when upon causes emergent and judicially discussed, and when upon responses and conferences touching general points or questions; how they are to be pressed, rigorously or tenderly; how they are to be mitigated by equity and good conscience, and whether discretion and strict law are to be mingled in the same courts, or kept apart in several courts; again, how the practice, profession, and erudition of law is to be censured and governed; and many other points touching the administration, and, as I may term it, animation of laws. Upon which I insist the less, because I purpose, if God give me leave (having begun a work of this nature in aphorisms), to propound it hereafter, noting it in the mean time for deficient.

50.   And for your Majesty's laws of England, I could say much of their dignity, and somewhat of their defect; but they cannot but excel the civil laws in fitness for the government: for the civil law was *non hos quaesitum munus in usus;*[421] it was not made for the countries which it governeth: hereof I cease to speak because I will not intermingle matter of action with matter of general learning.

XXIV   Thus have I concluded this portion of learning touching civil knowledge; and with civil knowledge have concluded human philosophy; and with human philosophy, philosophy in general. And being now at some pause, looking back into that I have passed through, this writing seemeth to me, *si nunquam fallit imago*[422] (as far as a man can judge of his own work), not much better than that noise or sound which musicians make while they are tuning their instruments: which is nothing pleasant to hear, but yet is a cause why the music is sweeter afterwards: so have I been content to tune the instruments of the Muses, that they may play that have better hands. And surely, when I set before me the condition of these times, in which learning hath made her third visitation or circuit in all the qualities thereof—as the excellency and vivacity of the wits of this age; the noble helps and lights which we have by the travails of ancient writers; the art of printing, which communicateth books to men of all fortunes; the openness of the world by navigation, which hath disclosed multitudes of experiments, and a mass of natural history; the leisure wherewith these times abound, not employing men so generally in civil business, as the states of Graecia did, in respect of their popularity, and the state of Rome, in respect of the greatness of their monarchy; the present disposition of these times at this instant to peace; the consumption of all that ever can be said in controversies of religion, which have so much diverted men from other sciences; the perfection of your Majesty's learning, which as a Phoenix may call whole vollies of wits to follow you; and the inseparable propriety of time, which is ever more and more to disclose truth—I cannot but be raised to this persuasion that this third period of time will far surpass that of the Grecian and Roman learning: only if men will know their own strength, and their own weakness both; and take, one from the other, light of invention, and not fire of contradiction; and esteem of the inquisition of truth as of an enterprise, and not as of a quality or ornament; and employ wit and magnificence to things of worth and excellency, and not to things vulgar and of popular estimation. As for my labours, if any man shall please himself or others in the reprehension of them, they shall make that ancient and patient request, *Verbera, sed audi;*[423] let men reprehend them, so they observe and weigh them: for the appeal is lawful, though it may be it shall not be needful, from the first cogitations of men to their second, and from the nearer times to the times further off. Now let us come to that learning, which both the former times were not so blessed as to know, sacred and inspired divinity, the Sabbath and port of all men's labours and peregrinations.

XXV

1. The prerogative of God extendeth as well to the reason as to the will of man; so that as we are to obey His law, though we find a reluctation in our will, so we are to believe His word, though we find a reluctation in our reason. For if we believe only that which is agreeable to our sense, we give consent to the matter, and not to the author; which is no more than we would do towards a suspected and discredited witness; but that faith which was accounted to Abraham for righteousness was of such a point as whereat Sarah laughed,[424] who therein was an image of natural reason.

2. Howbeit, if we will truly consider it, more worthy it is to believe than to know as we now know. For in knowledge man's mind suffereth from sense; but in belief it suffereth from spirit, such one as it holdeth for more authorized than itself, and so suffereth from the worthier agent. Otherwise it is of the state of man glorified; for then faith shall cease, and we shall know as we are known.

3. Wherefore we conclude that sacred theology (which in our idiom we call divinity) is grounded only upon the word and oracle of God, and not upon the light of nature: for it is written, *Caeli enarrant gloriam Dei;*[425] but it is not written, *Caeli enarrant voluntatem Dei:* but of that it is said, *Ad legem et testimonium: si non fecerint secundum verbum istud,*[426] etc. This holdeth not only in those points of faith which concern the mysteries of the Deity, of the Creation, of the Redemption, but likewise those which concern the moral law truly interpreted: *Love your enemies: do good to them that hate you; be like to your heavenly Father, that suffereth his rain to fall upon the just and unjust.*[427] To this it ought to be applauded, *nec vox hominem sonat:*[428] it is a voice beyond the light of nature. So we see the heathen poets, when they fall upon a libertine passion, do still expostulate with laws and moralities, as if

they were opposite and malignant to nature; *Et quod natura remittit, invida jura negant.*[429] So said Dendamis the Indian unto Alexander's messengers, that he had heard somewhat of Pythagoras, and some other of the wise men of Graecia, and that he held them for excellent men: but that they had a fault, which was that they had in too great reverence and veneration a thing which they called law and manners.[430] So it must be confessed, that a great part of the law moral is of that perfection, whereunto the light of nature cannot aspire: how then is it that man is said to have, by the light and law of nature, some notions and conceits of virtue and vice, justice and wrong, good and evil? Thus, because the light of nature is used in two several senses; the one, that which springeth from reason, sense, induction, argument, according to the laws of heaven and earth; the other, that which is imprinted upon the spirit of man by an inward instinct, according to the law of conscience, which is a sparkle of the purity of his first estate; in which latter sense only he is participant of some light and discerning touching the perfection of the moral law: but how? sufficient to check the vice, but not to inform the duty. So then the doctrine of religion, as well moral as mystical, is not to be attained but by inspiration and revelation from God.

4.   The use notwithstanding of reason in spiritual things, and the latitude thereof, is very great and general: for it is not for nothing that the apostle calleth religion *our reasonable service of God;*[431] insomuch as the very ceremonies and figures of the old law were full of reason and signification, much more than the ceremonies of idolatry and magic, that are full of non-significants and surd characters. But most especially the Christian faith, as in all things, so in this deserveth to be highly magnified; holding and preserving the golden mediocrity in this point between the law of the heathen and the law of Mahomet, which have embraced the two extremes. For the religion of the heathen had no constant belief or confession, but left all to the liberty of argument; and the religion of Mahomet, on the other side, interdicteth argument altogether: the one having the very face of error, and the other of imposture: whereas the faith doth both admit and reject disputation with difference.

5.   The use of human reason in religion is of two sorts: the former,

in the conception and apprehension of the mysteries of God to us revealed; the other, in the inferring and deriving of doctrine and direction thereupon. The former extendeth to the mysteries themselves; but how? by way of illustration, and not by way of argument: the latter consisteth indeed of probation and argument. In the former, we see, God vouchsafeth to descend to our capacity, in the expressing of his mysteries in sort as may be sensible unto us; and doth graft his revelations and holy doctrine upon the notions of our reason, and applieth his inspirations to open our understanding, as the form of the key to the ward of the lock: for the latter, there is allowed us a use of reason and argument, secondary and respective, although not original and absolute. For after the articles and principles of religion are placed and exempted from examination of reason, it is then permitted unto us to make derivations and inferences from and according to the analogy of them, for our better direction. In nature this holdeth not; for both the principles are examinable by induction, though not by a medium or syllogism; and besides, those principles or first positions have no discordance with that reason which draweth down and deduceth the inferior positions. But yet it holdeth not in religion alone, but in many knowledges, both of greater and smaller nature, namely, wherein there are not only *posita* but *placita;* for in such there can be no use of absolute reason. We see it familiarly in games of wit, as chess, or the like: the draughts and first laws of the game are positive, but how? merely *ad placitum,* and not examinable by reason; but then how to direct our play thereupon with best advantage to win the game, is artificial and rational. So in human laws, there be many grounds and maxims which are *placita juris,* positive upon authority, and not upon reason, and therefore not to be disputed: but what is most just, not absolutely but relatively, and according to those maxims, that affordeth a long field of disputation. Such therefore is that secondary reason, which hath place in divinity, which is grounded upon the *placets* of God.

6.   Here therefore I note this deficiency, that there hath not been, to my understanding, sufficiently inquired and handled the true limits and use of reason in spiritual things, as a kind of divine dialectic: which for that it is not done, it seemeth to me a thing usual, by pretext of true conceiving that which is revealed, to search and mine into that which

is not revealed; and by pretext of enucleating inferences and contradictories, to examine that which is positive: the one sort falling into
the error of Nicodemus, demanding to have things made more sensible
than it pleaseth God to reveal them, *Quomodo possit homo nasci cum
sit senex?*[432] the other sort into the error of the disciples, which were
scandalized at a show of contradiction, *Quid est hoc quod dicit nobis?
Modicum, et non videbitis me; et iterum modicum, et videbitis me,* etc.[433]

7.   Upon this I have insisted the more, in regard of the great and blessed
use thereof; for this point, well laboured and defined of, would in my
judgment be an opiate to stay and bridle not only the vanity of curious
speculations, wherewith the schools labour, but the fury of controversies,
wherewith the church laboureth. For it cannot but open men's eyes,
to see that many controversies do merely pertain to that which is either
not revealed, or positive; and that many others do grow upon weak
and obscure inferences or derivations: which latter sort, if men would
revive the blessed style of that great doctor of the Gentiles, would be
carried thus, *ego, non dominus;*[434] and again, *secundum consilium meum,*
in opinions and counsels, and not in positions and oppositions. But
men are now overready to usurp the style, *non ego, sed dominus;* and
not so only, but to bind it with the thunder and denunciation of curses
and anathemas, to the terror of those which have not sufficiently learned
out of Salomon, that *the causeless curse shall not come.*[435]

8.   Divinity hath two principal parts; the matter informed or revealed,
and the nature of the information or revelation: and with the latter we
will begin, because it hath most coherence with that which we have
now last handled. The nature of the information consisteth of three
branches; the limits of the information, the sufficiency of the information, and the acquiring or obtaining the information. Unto the limits
of the information belong these considerations; how far forth particular
persons continue to be inspired; how far forth the Church is inspired;
how far forth reason may be used: the last point whereof I have noted
as deficient. Unto the sufficiency of the information belong two considerations; what points of religion are fundamental, and what perfective, being matter of further building and perfection upon one and the
same foundation; and again, how the gradations of light, according to
the dispensation of times, are material to the sufficiency of belief.

9. Here again I may rather give it in advice, than note it as deficient, that the points fundamental, and the points of further perfection only, ought to be with piety and wisdom distinguished: a subject tending to much like end as that I noted before; for as that other were likely to abate the number of controversies, so this is like to abate the heat of many of them. We see Moses when he saw the Israelite and the Egyptian fight, he did not say, *Why strive you?* but drew his sword and slew the Egyptian: but when he saw the two Israelites fight, he said, *You are brethren, why strive you?*[436] If the point of doctrine be an Egyptian, it must be slain by the sword of the spirit, and not reconciled; but if it be an Israelite, though in the wrong, then, *Why strive you?* We see of the fundamental points, our Saviour penneth the league thus, *He that is not with us, is against us;*[437] but of points not fundamental, thus, *He that is not against us, is with us.*[438] So we see the coat of our Saviour was entire without seam,[439] and so is the doctrine of the Scriptures in itself; but the garment of the Church was of divers colours,[440] and yet not divided: we see the chaff may and ought to be severed from the corn in the ear, but the tares may not be pulled up from the corn in the field.[441] So as it is a thing of great use well to define what, and of what latitude those points are, which do make men merely aliens and disincorporate from the Church of God.

10. For the obtaining of the information, it resteth upon the true and sound interpretation of the Scriptures, which are the fountains of the water of life. The interpretations of the Scriptures are of two sorts; methodical, and solute or at large. For this divine water,[442] which excelleth so much that of Jacob's Well, is drawn forth much in the same kind as natural water useth to be out of wells and fountains; either it is first forced up into a cistern, and from thence fetched and derived for use; or else it is drawn and received in buckets and vessels immediately where it springeth. The former sort whereof, though it seem to be the more ready, yet in my judgment is more subject to corrupt. This is that method which hath exhibited unto us the scholastical divinity; whereby divinity hath been reduced into an art, as into a cistern, and the streams of doctrine or positions fetched and derived from thence.

11. In this men have sought three things, a summary brevity, a compacted strength, and a complete perfection; whereof the two first they fail to find, and the last they ought not to seek. For as to brevity, we see in all summary methods, while men purpose to abridge, they give cause to dilate. For the sum or abridgment by contraction becometh obscure; the obscurity requireth exposition, and the exposition is deduced into large commentaries, or into common places and titles, which grow to be more vast than the original writings, whence the sum was at first extracted. So, we see, the volumes of the schoolmen are greater much than the first writings of the fathers, whence the Master of the Sentences[443] made his sum or collection. So, in like manner, the volumes of the modern doctors of the civil law exceed those of the ancient jurisconsults, of which Tribonian[444] compileth the digest. So as this course of sums and commentaries is that which doth infallibly make the body of sciences more immense in quantity, and more base in substance.

12.   And for strength, it is true that knowledges reduced into exact methods have a show of strength, in that each part seemeth to support and sustain the other; but this is more satisfactory than substantial: like unto buildings which stand by architecture and compaction, which are more subject to ruin than those which are built more strong in their several parts, though less compacted. But it is plain that the more you recede from your grounds, the weaker do you conclude: and as in nature, the more you remove yourself from particulars, the greater peril of error you do incur: so much more in divinity, the more you recede from the Scriptures by inferences and consequences, the more weak and dilute are your positions.

13.   And as for perfection or completeness in divinity, it is not to be sought; which makes this course of artificial divinity the more suspect. For he that will reduce a knowledge into an art, will make it round and uniform: but in divinity many things must be left abrupt, and concluded with this: *O altitudo sapientiae et scientiae Dei! quam incomprehensibilia sunt judicia ejus, et non investigabiles viae ejus!*[445] So again the apostle saith, *Ex parte scimus:*[446] and to have the form of a total, where there is but matter for a part, cannot be without supplies by supposition and presumption. And therefore I conclude, that the true use of these sums and methods hath place in institutions or introductions preparatory unto knowledge: but in them, or by deducement

from them, to handle the main body and substance of a knowledge, is in all sciences prejudicial, and in divinity dangerous.

14. As to the interpretation of the Scriptures solute and at large, there have been divers kinds introduced and devised; some of them rather curious and unsafe than sober and warranted. Notwithstanding, thus much must be confessed, that the Scriptures being given by inspiration, and not by human reason, do differ from all other books in the author: which, by consequence, doth draw on some difference to be used by the expositor. For the inditer of them did know four things which no man attains to know; which are, the mysteries of the kingdom of glory, the perfection of the laws of nature, the secrets of the heart of man, and the future succession of all ages. For as to the first it is said, *He that presseth into the light, shall be oppressed of the glory.* And again, *No man shall see my face and live.*[447] To the second, *When he prepared the heavens I was present, when by law and compass he inclosed the deep.*[448] To the third, *Neither was it needful that any should bear witness to him of man, for he knew well what was in man.*[449] And to the last, *From the beginning are known to the Lord all his works.*[450]

15. From the former two of these have been drawn certain senses and expositions of Scriptures, which had need be contained within the bounds of sobriety; the one anagogical, and the other philosophical. But as to the former, man is not to prevent his time: *Videmus nunc per speculum in aenigmate, tunc autem facie ad faciem:*[451] wherein nevertheless there seemeth to be a liberty granted, as far forth as the polishing of this glass, or some moderate explication to this enigma. But to press too far into it, cannot but cause a dissolution and overthrow of the spirit of man. For in the body there are three degrees of that we receive into it, aliment, medicine, and poison; whereof aliment is that which the nature of man can perfectly alter and overcome: medicine is that which is partly converted by nature, and partly converteth nature; and poison is that which worketh wholly upon nature, without that, that nature can in any part work upon it. So in the mind, whatsoever knowledge reason cannot at all work upon and convert is a mere intoxication, and endangereth a dissolution of the mind and understanding.

16. But for the latter, it hath been extremely set on foot of late time by the school of Paracelsus, and some others, that have pretended to find the truth of all natural philosophy in the Scriptures; scandalizing and traducing all other philosophy as heathenish and profane. But there is no such enmity between God's word and His works; neither do they give honour to the Scriptures, as they suppose, but much imbase them. For to seek heaven and earth in the word of God (whereof it is said, *Heaven and earth shall pass, but my word shall not pass*[452]) is to seek temporary things amongst eternal: and as to seek divinity in philosophy is to seek the living amongst the dead,[453] so to seek philosophy in divinity is to seek the dead amongst the living: neither are the pots or lavers, whose place was in the outward part of the temple, to be sought in the holiest place of all, where the ark of the testimony was seated. And again, the scope or purpose of the spirit of God is not to express matters of nature in the Scriptures, otherwise than in passage, and for application to man's capacity, and to matters moral or divine. And it is a true rule, *auctoris aliud agentis parva auctoritas;* for it were a strange conclusion, if a man should use a similitude for ornament or illustration sake, borrowed from nature or history according to vulgar conceit, as of a Basilisk, an Unicorn, a Centaur, a Briareus, a Hydra, or the like, that therefore he must needs be thought to affirm the matter thereof positively to be true. To conclude, therefore, these two interpretations, the one by reduction or enigmatical, the other philosophical or physical, which have been received and pursued in imitation of the rabbins and cabalists,[454] are to be confined with a *noli altum sapere, sed time.*[455]

17.   But the two latter points, known to God and unknown to man, touching the secrets of the heart, and the successions of time, doth make a just and sound difference between the manner of the exposition of the Scriptures and all other books. For it is an excellent observation which hath been made upon the answers of our Saviour Christ to many of the questions which were propounded to him, how that they are impertinent to the state of the question demanded; the reason whereof is, because not being like man, which knows man's thoughts by his words, but knowing man's thoughts immediately, he never answered their words, but their thoughts: much in the like manner it is with the Scriptures, which being written to the thoughts of men, and to the succession of all ages, with a foresight of all heresies, contradictions,

differing estates of the church, yea and particularly of the elect, are not to be interpreted only according to the latitude of the proper sense of the place, and respectively towards that present occasion whereupon the words were uttered, or in precise congruity or contexture with the words before or after, or in contemplation of the principal scope of the place; but have in themselves, not only totally or collectively, but distributively in clauses and words, infinite springs and streams of doctrine to water the church in every part. And therefore as the literal sense is, as it were, the main stream or river; so the moral sense chiefly, and sometimes the allegorical or typical, are they whereof the church hath most use; not that I wish men to be bold in allegories, or indulgent or light in allusions: but that I do much condemn that interpretation of the Scripture which is only after the manner as men use to interpret a profane book.

18. In this part, touching the exposition of the Scriptures, I can report no deficience; but by way of remembrance this I will add: in perusing books of divinity, I find many books of controversies; and many of commonplaces and treaties; a mass of positive divinity, as it is made an art; a number of sermons and lectures, and many prolix commentaries upon the Scriptures, with harmonies and concordances: but that form of writing in divinity which in my judgment is of all others most rich and precious, is positive divinity, collected upon particular texts of Scriptures in brief observations; not dilated into commonplaces, not chasing after controversies, not reduced into method of art; a thing abounding in sermons, which will vanish, but defective in books which will remain; and a thing wherein this age excelleth. For I am persuaded (and I may speak it with an *absit invidia verbo*,[456] and no ways in derogation of antiquity, but as in a good emulation between the vine and the olive) that if the choice and best of those observations upon texts of Scriptures, which have been made dispersedly in Sermons within this your Majesty's island of Britain by the space of these forty years and more, leaving out the largeness of exhortations and applications thereupon, had been set down in a continuance, it had been the best work in divinity which had been written since the Apostles' times.

19. The matter informed by divinity is of two kinds; matter of belief and truth of opinion, and matter of service and adoration; which is also judged and directed by the former: the one being as the internal soul of religion, and the other as the external body thereof. And therefore the heathen religion was not only a worship of idols, but the whole religion was an idol in itself; for it had no soul, that is, no certainty of belief or confession: as a man may well think, considering the chief doctors of their church were the poets: and the reason was, because the heathen gods were no jealous gods, but were glad to be admitted into part, as they had reason. Neither did they respect the pureness of heart, so they might have external honour and rites.

20.   But out of these two do result and issue four main branches of divinity; *faith, manners, liturgy,* and *government.* Faith containeth the doctrine of the nature of God, of the attributes of God, and of the works of God. The nature of God consisteth of three persons in unity of Godhead. The attributes of God are either common to the Deity, or respective to the persons. The works of God summary are two, that of the *creation* and that of the *redemption:* and both these works, as in total they appertain to the unity of the Godhead, so in their parts they refer to the three persons: that of the creation, in the mass of the matter, to the Father; in the disposition of the form, to the Son; and in the continuance and conservation of the being, to the Holy Spirit. So that of the redemption, in the election and counsel, to the Father; in the whole act and consummation to the Son; and in the application, to the Holy Spirit; for by the Holy Ghost was Christ conceived in flesh, and by the Holy Ghost are the elect regenerate in spirit. This work likewise we consider either effectually, in the elect; or privatively in the reprobate; or according to appearance, in the visible church.

21.   For manners, the doctrine thereof is contained in the law, which discloseth sin. The law itself is divided, according to the edition thereof, into the law of nature, the law moral, and the law positive; and according to the style, into negative and affirmative, prohibitions and commandments. Sin, in the matter and subject thereof, is divided according to the commandments; in the form thereof, it referreth to the three persons in Deity: sins of infirmity against the Father, whose more special attribute is power; sins of ignorance against the Son, whose attribute is wisdom; and sins of malice against the Holy Ghost, whose attribute is grace or love. In the motions of it, it either moveth to the right hand or to the left; either to blind devotion, or to profane and libertine transgression; either in imposing restraint where God granteth liberty, or in taking liberty where God imposeth restraint. In the degrees and progress of it, it divideth itself into thought, word, or act. And in this part I commend much the deducing of the law of God to cases of conscience; for that I take indeed to be a breaking, and not exhibiting whole of the bread of life. But that which quickeneth both these doctrines of faith and manners, is the elevation and consent of the heart; whereunto appertain books of exhortation, holy meditation, Christian resolution, and the like.

22.   For the liturgy or service, it consisteth of the reciprocal acts between God and man; which, on the part of God, are the preaching of the word, and the sacraments, which are seals to the covenant, or as the visible word; and on the part of man, invocation of the name of God; and under the law, sacrifices; which were as visible prayers or confessions: but now the adoration being *in spiritu et veritate,*[457] there remaineth only *vituli labiorum;*[458] although the use of holy vows of thankfulness and retribution may be accounted also as sealed petitions.

23.   And for the government of the church, it consisteth of the patrimony of the church, the franchises of the church, and the offices and jurisdictions of the church, and the laws of the church directing the whole; all which have two considerations, the one in themselves, the other how they stand compatible and agreeable to the civil estate.

24.   This matter of divinity is handled either in form of instruction of truth, or in form of confutation of falsehood. The declinations from religion, besides the privative, which is atheism and the branches thereof, are three; *Heresies, Idolatry,* and *Witchcraft;* heresies, when we serve the true God with a false worship; idolatry, when we worship false gods, supposing them to be true; and witchcraft, when we adore false gods, knowing them to be wicked and false: for so your Majesty doth excellently well observe, that witchcraft is the height of idolatry. And yet we see though these be true degrees, Samuel teacheth us that they are all of a nature, when there is once a receding from the word of God; for so he saith, *Quasi peccatum ariolandi est repugnare, et quasi scelus idololatriae nolle acquiescere.*[459]

25.   These things I have passed over so briefly because I can report no deficience concerning them: for I can find no space or ground that lieth vacant and unsown in the matter of divinity: so diligent have men been, either in sowing of good seed, or in sowing of tares.

Thus have I made as it were a small globe of the intellectual world, as truly and faithfully as I could discover: with a note and description of those parts which seem to me not constantly occupate, or not well converted by the labour of man. In which, if I have in any point receded from that which is commonly received, it hath been with a purpose of proceeding *in melius,* and not *in aliud;* a mind of amendment and proficience, and not of change and difference. For I could not be true and constant to the argument I handle, if I were not willing to go beyond others; but yet not more willing than to have others go beyond me again: which may the better appear by this, that I have propounded my opinions naked and unarmed, not seeking to preoccupate the liberty of men's judgments by confutations. For in anything which is well set down, I am in good hope, that if the first reading move an objection, the second reading will make an answer. And in those things wherein I have erred, I am sure I have not prejudiced the right by litigious arguments; which certainly have this contrary effect and operation, that they add authority to error, and destroy the authority of that which is well invented: for question is an honour and preferment to falsehood, as on the other side it is a repulse to truth. But the errors I claim and challenge to myself as mine own: the good, if any be, is due *tanquam adepts sacrificii,*[460] to be incensed to the honour, first of the Divine Majesty, and next of your Majesty, to whom on earth I am most bounden.

DEO GLORIA

NOTES ॐ

THE FIRST BOOK

1. Prov. 25:3.
2. *Phaedo* 1.72.
3. 1 Kings 4:29.
4. Tac. *Annal.* 13.3.
5. Marsilius Ficinus *Arg. ad Herm. Trism.*—Et philosophus maximus, et sacerdos maximus, et rex maximus.
6. *Propriety inherent;* the logical "Proprium quod consequitur essentiam rei."
7. 1 Cor. 8:1.
8. Eccles. 12:12.
9. Eccles. 1:18.
10. Col. 2:8.
11. See Gen. 2 and 3.
12. Eccles. 1:8.
13. Eccles. 3:11.
14. Prov. 20:27.
15. 1 Cor. 13:1.
16. Eccles. 2:13, 14.
17. ὁ σκοτεινός.

  Μὴ ταχὺς Ἡρακλείτου ἐπ' ὀμφαλὸν εἴλεο βίβλον·
    Τοῦ 'φεσίου μάλα τοι δύσβατος ἀτραπιτός·
  Ὀρφνη καὶ σκότος ἐστὶν ἀλάμπετον, ἢν δέ ϲε μύϲ ϲ ϳϳς·
    Εἰσαγάγῃ, φανεροῦ λαμπρότερ' ἠελίου.
                              Diog. Laert. 9.

18. Αὐγὴ ξηρὴ ψυχὴ σοφωτάτη. A corruption of αὔη ψυχὴ σοφωτάτη. (See note in Ellis and Spedding's edition.) The phrase occurs in Stobaeus, cf. Ritter *Hist. Philos.* vol. 1, *Heraclitus.*
19. Philo Jud. *de Somn.*
20. See Bacon's *Essays—On Atheism.*
21. Job 13:7.
22. Hom. *Il.* 8.19.
23. See Pliny *Nat. Hist.* 7.31.
24. Plut. *vit. Cat.*
25. Virg. *Aen.* 6.851.
26. Plato *Apol. Soc.* 1.19, 24.
27. Cf. Aristotle *Rhet.* 2.14.4, where he says that the body reaches perfection at the age of 35 (7 x 5), and the mind at 49 (7 x 7).
28. See Plato *Rep.* 5.473.
29. Guicciard. 16.5.
30. Pind. *Pyth.* 2.21, seq.
31. Cic. *ad Att.* 2.1.

32. Seneca *Epist.* 3, quoted from Pomponius, "Quidam adeo in latebras refugerunt, ut" etc.

33. Plutarch. Libanius *Vit. Demosth.* (Edition Dindorf, p. 6.) Told of Pytheas, not of Aeschines.

34. Hor. *Odes* 3.2.

35. Mach. *Disc. sopra Tit. Liv.* 3.1, speaking of the Franciscan and Dominican orders.

36. Livii *Praef.*

37. Sallust *Ep.* 1. *De Rep. ord.*

38. Diog. Laert. 6.54.

39. Prov. 28:22.

40. Prov. 23:23.

41. Tac. *Ann.* 3.76, ad fin.

42. Joel 2:28.

43. Conference of Agesilaus and Pharnabazus. Plut. *Vit. Ages.*

44. Ovid *Ep.* 15.83.

45. *De Augm.* has *nullum occurrit dedecus literis ex litteratorum moribus, quatenus, sunt literati, adhaerens,* which explains it. The *not* before *inherent* goes with *cannot* according to the rule of double negative, as it prevailed in early English writers.

46. Plutarch *Vit. Solon.*

47. Plato *Epist.* Z.3.331.

48. Sall. *Epist. de Rep. ord.*

49. Cic. *ad Att.* 2.1.

50. Cic. *pro Mur.* 31.65.

51. Ovid *A. Am.* 2.548.

52. Demosth. *Chers.* 187, ad finem.

53. The *Quinquennium Neronis* refers to the first five years of Nero's reign, during which he was under Seneca's influence.

54. Matt. 25:20.

55. A saying of Epicurus. Seneca *Epist. Mor.* 1.7.

56. Herod. 1.99.

57. Prov. 25:3.

58. Plutarch *Vit. Themist.,* ad init.

59. Plat. *Conv.* 3.215, where the thought is present, though the exact similitude is wanting.

60. Lucian. *de Merc. Cond.* 33, 34.

61. See *Bethulian's Rescue,* bk. 5.
   "Tous ces esprits dont la voix flattereuse
   Change Hécube en Hélène, et Faustine en Lucrèce."

62. Diog. Laert. *Vit. Aristippi* 2.69; the answer was given by Aristippus.

63. Ibid. 2.79.

64. Spartianus *Vit. Adriani* § 15. The excuse was made by Favorinus.

65. Had Bacon been accustomed to use the then modern word *its,* it is probable he would have used it here. As it is "the state and virtue" must mean its pure and right condition.

66. *Discourse of reason:* process of reasoning. Cf. Sanderson *Ars Log.* 3.1.

67. John 7:10.

68. Bishop of Silves, died 1580.

69. *Colloq.* between *Juvenis* and *Echo.*

70. Ovid *Metam.* 10.243.

71. Theocr. 5.2 (schol.) or Erasmi *Adag.*

72. 1 Tim. 6:20.

73. For his judgment—a harsh one—on the Schoolmen, see the *Nov. Org.* 1.71.

74. See Hallam *Hist. of Lit.* vol 1. init. § 18–23.

75. Rerum pondera minutissimis sententiis fregit. Quint. *de Inst. Orat.* 10.1.

76. Virg. *Ecl.* 6.75.

77. Diog. Laert. 3.18 *(Vit. Platonis).*

78. Hor. *Ep.* 1.18.69.

79. Tac. *Hist.* 1.51.

80. Cardan—born in Pavia, 1501—wrote about 122 works on Physics, Mathematics, Astronomy, Astrology, Medicine, Ethics, Music, etc.

81. Albertus Magnus—born in Swabia, about 1198—the most learned man of his age.

82. Οαυμάσια Ἀκούσματα—a treatise now recognised as spurious.

83. Arist. *Soph. El.* 2.

84. Jer. 6:16.

85. See *Nov. Org.* 1.84.

86. Liv. 9.17.

87. Sext. Empir. *adv. Math.* 7.133.

88. See *Nov. Org.* 1.63.

89. See *Nov. Org.* 1.64.

90. *Tuscul. Disp* 1.10.20. He is speaking of Aristoxenus. Plato, in the *Phaedo,* introduces the same analogy.

91. *De Gener. et Corrupt.* 1.2.

92. Cic. *De Nat. Deor.* 1.8.18.

93. His Εἰρώνεια. See Plato *Apol.* for the best instance of this. He there explains his superiority to consist in the knowledge of his own ignorance.

94. Ovid *Metam.* 10.667.

95. Cic. *Tusc. Disp.* 5.4.10.

96. In all editions *hath.* For in Bacon's time the verb singular was very commonly used with more nominatives than one, and even with plural nouns, as here.

97. Prov. 27:6.

98. Cf. Hooker *Ecclesiastical Polity* 1.4.1, 2.

99. Dionys. *De Coelesti Hierarch.* cap. 7, 8, 9. This work is, as Bacon hints, spurious, though no other author is assigned.

100. Gen. 1:3.

101. Gen. 2:3.

102. Gen. 2:19.

103. Gen. 4:2.

104. Gen. 4:21, 22.

105. Gen. 11.

106. Act. Ap. 7.22.

107. Plat. *Tim.* 3:22.

108. Levit. 13:12–14.

109. Job 26:7.

110. Job 26:13.

111. Job 38:31.

112. Job 9:9.

113. Job 10:10.

114. Job 28:1.

115. 1 Kings 3:5, seq.

116. *Nov. Org.* 2.30.

117. 1 Kings 4:33.

118. Prov. 25:2.

119. Luke 2:46.

120. Act. Ap. 2.1.

121. Gibbon vol. 2. c. 23, who quotes Ammian. 25.5.

122. Gibbon vol. 4. c. 45. The story that St. Gregory destroyed the Palatine Library is now rejected; but as to his aversion to profane letters there can be no doubt. Milman's *Latin Christianity* bk. 3. c. 7.

123. Pss. 19, 104.

124. Matt. 22:29.

125. Cf. *Nov. Org.* 1.89.

126. Plat. *Rep.* 5.473.

127. Hor. *Odes* 2.10.19.

128. *Agric. Vit.* c. 3.

129. Τίσειαν Δαναοὶ ἐμὰ δάκρυα σοῖσι βέλεσσιν. Hom. *Il.* 1.42. Dionys. *Epit.* (Xiphilini) 12.

130. Matt. 10:41.

131. See Dante *Purgatorio* 10, who seems to take it from the *Life of Gregory*, by John the Deacon.

132. C. Plin. *Epist.* 10.97.

133. Plutarch *Apophth.* 179.

134. Unum de istis puto qui cuminum secant. Julian *Caes.* So Aristot. *Eth. Nic.* 4.3, εἷς τῶν διαπρίοντων τὸ κύμινον, where, however, the phrase is used of the "skinflint," or niggard.

135. Acts 26:28.

136. Better known as L. Aurelius Verus.

137. See his life by Spartianus.

138. These anecdotes of Alexander come from Plutarch *Vit. Alex.*

139. Sen. *De Benef.* 5.4.

140. Sen. *Ep. Mor.* 6.7.

141. Ἰχώρ, οἷός πέρ τε ῥέει μακάρεσσι θεοῖσι. *Il.* 5.340. Cf. Seneca *Ep. ad Lucil.* 59.

142. The Greek is λευκοπάρυφος.

143. ὁλοπόρφυρος. *Apophth. Reg. et Imp.*

144. Cic. *Brutus* 72.

145. Cic. *ad. Att.* 12.40, 41; 13.50.

146. Cic. *Epist. ad Div.* 9.16.

147. Eccles. 12:11.

148. Suet. *Jul. Caes.* c. 70.

149. Ibid.

150. Cf. Hor. *Sat.* 1.7.

151. Plutarch *Jul. Caes.*

152. To these might have well been added Caesar's exhortation to the boatman, "Thou carriest Caesar and his fortunes."

153. Suet. *Jul. Caes.* c. 77.

154. Xen. *Anab.* 2, ad fin.

155. Xen. *Anab.* 2.1.12.

156. Ov. *Ep. ex Pont.* 2.9.47.

157. Eccles. 1:9.

158. Ἔοικεν, ὦ ἄνδρες, ὅτε Δαρεῖον ἡμεῖς ἐνικῶμεν ἐνταῦθα, ἐκεῖτις ἐν Ἀρκαδίᾳ γεγονέναι μυομαχία. Plut. *Ages.* c. 15.

159. There is no such tale in Epictetus, but see *Simplicii in Epict. Comment.* cap. 33.

160. Virg. *Georg.* 2.490.

161. Xen. *Mem.* 1.6.

162. *Georg.* 4.561, 562.

163. Rev. 2:24.

164. Cf. Herod. 2.141 for the ascendency of the Priesthood in Egypt.

165. *De Rer. Nat.* 2.1–10.

166. The Latin has *Nos autem . . . conculcantes haec rudimenta . . . novimus.* Perhaps *in* should be omitted—"do disclaim these rudiments of the senses."

167. Ovid *Metam.* 11.153, seq.

168. Tacit. *Annal.* 14.9.

169. Cf. Cic. *de Orat.* 1.44, where it is *Ithaca,* not his old wife, that Ulysses is said to prefer to immortality.

170. Matt. 11:19.

THE SECOND BOOK

1. A favourite thought of Bacon's, and expressed afterwards on the engraved title page of the first edition of the *Novum Organum,* A.D. 1620.

2. Eccles. 10:10.

3. Virg. *Georg.* 4.8.

4. *Orat. post Redit. in Sen.* 12.30, which in Bacon's day was counted genuine. The actual passage is something stronger, for it has *nefas* instead of *ingratum.*

5. Phil. 3:13.

6. Liv. 2.32.

7. 1 Sam. 30:22.

8. Virg. *Georg.* 3.128.

9. See *Nov. Org.* 2.7: "Transeundum plane a Vulcano ad Minervam, si in animo sit veras corporum texturas et schematismos . . . in lucem protrahere."

10. Aelian *Var. Hist.* 4.19 says that Philip held him, and Athenaeus 9.398 f. states the amount said to have been allowed him by Alexander, 800 talents. But Bacon takes his statement here from Plin. *Nat. Hist.* 8.17.

11. The Latin has for "travail in arts of Nature," "*in labyrinthis artium viam sibi aperiunt,*"—where Art is opposed to Nature. So that the phrase "Arts of Nature" must be modified to mean "Arts concerned with Nature." Or, possibly, there is some mistake in the reading. All the old editions have *travailes.* If the reading is correct, the sense will be that they who lay down rules and general principles of Arts in things Natural are worthy of higher reward than are they who only collect Histories, i.e. catalogues or registers of detached facts.

12. *Sylva:* de Orat. 3.26 (103). *Supellex:* Orat. 24 (80).

13. Cic. *ad Att.* 9.7.10.

14. James 1:17.

15. Exod. 7.10. It was *Aaron's* rod that became a serpent.

16. Publ. Syr. *Sentent.* 166: Amare et sapere vix Deo conceditur.

17. Ennius, quoted by Cic. *de Off.* 1.16 (5).

18. Prov. 22:13.

19. Virg. *Aen.* 5.231.

20. Cf. *Nov. Org.* 1.45 and 2.28. These "instances of exception to general kinds" he there terms *instantiae monodicae,* quas etiam *irregulares* sive *heteroclitas* appellare consuevimus.

21. *De Mirabilibus Auscultationibus (θαυμάσια ἀκούσματα);* see book one, note 82 (p. 231).

22. *Mirabilaries.* In *De Augm. Sc.* 2, he calls them "Mirabilarix et prodigiastri."

23. Cf. *Nov. Org.* 1.120. This thought is to be met with in Chaucer, *Persone's Tale:* "Certes, Holy Writ may not be defouled, no more than the sonne that shineth on the myxene."

24. Plato *Hipp. Maj.* 3.288, 291.

25. Thales. See Plat. *Theaet.* 1.174.

26. Aristot. *Polit.* 1.3.1 and *Phys.* 1.

27. Virg. *Georg.* 4.387, sqq.

28. Cf. *Nov. Org.* 1.77.

29. *Story* here = history: "librorum neutiquam historicorum."

30. As in the Epitomes written in the decline of Latin literature.

31. Job 26:7. "Qui appendit terram super nihilum."

32. Virg. *Aen.* 4.177.

33. Said of Cassius, Tac. *Ann.* 4.34: "Cremutius Cordus postulatur, . . . quod C. Cassium *Romanorum ultimum* dixisset." Cf. Plut. *Brutus* 43. Suet. *Tib.* 61, who attributes it to both Brutus and Cassius.

34. Cic. *Off.* 1.34.

35. Buchanan, for whom King James had no love.

36. Virg. *Ecl.* 1.67.

37. Virg. *Aen.* 3.96.

38. Ariosto *Orlando Furioso* end of bk. 34 and opening of bk. 35. (See Ellis and Spedding's edition of the *De Augm. Sc.*)

39. Virg. *Aen.* 5.751.

40. Plin. *Ep.* 3.21: "Postquam desiimus facere laudanda, laudari quoque ineptum putamus." Were Bacon's quotations usually from memory?

41. Prov. 10:7.

42. Cic. *Philip.* 9: "Vita enim mortuorum in memoria vivorum est posita." From Dem. *adv. Lept.* 488, ἵν' ἦν ζῶντες ἐκτήσαντο εὐδοξίαν αὕτη καὶ τελευτκόσιν αὐτοῖς ἀποδοθείη.

43. Tac. *Ann.* 13.31.

44. Esther 6:1.

45. See Plutarch *Sympos.* 1. Qu. 6.

46. Such books as Machiavelli's *Discorsi sopra Livio* are here meant.

47. Virg. *Georg.* 1.250, 251.

48. *Plus ultra* was the motto of Charles V. (Ellis).

49. Virg. *Aen.* 6.590.

50. Dan. 12:4.

51. 2 Pet. 3:8.

52. Hab. 2.2, but misquoted. "That he may run that readeth,"—i.e., may hasten to carry on the tidings.

53. Vid. Cic. *ad Fam.* 9.16.

54. Hor. *Ars Poetica* 9.

55. Virg. *Aen.* 4.178–180.

56. Not Pallas, but Thetis, Hom. *Il.* 1.401, sqq.

57. Hom. *Il.* 1.831 and Machiav. *Prince* c. 18.

58. In the Latin, in room of these examples, the fables of Pan, Perseus, and Diony-sus are expounded to show respectively how physical, political, and moral doctrines might be thence deduced.

59. Rather the *second* than the *third* part of learning—History, Poesy, Philosophy.

60. Plat. *Theaet.* 1.152. Ovid *Metam.* 15.165.

61. Eccles. 42:21.

62. Discourse on Livy 3.1.

63. See *Nov. Org.* 2.27. "Instantiae conformes."

64. Virg. *Aen.* 7.9.

65. Virg. *Aen.* 6.787.

66. Μικρόκοσμος—a favourite dogma with Paracelsus, who divided the body of man according to the cardinal points of the world. But Bacon is perhaps referring to the Platonists in the first part of the sentence.

67. Ps. 8:3.

68. Luke 20:25.

69. Hom. *Il.* 8.19–22.

70. Col. 2:18.

71. 2 Cor. 2:11.

72. The nature of Angels was a favourite subject of speculation and discussion among the Schoolmen, whose writings on it deserve Bacon's censure.

73. ἐν βυθῷ γὰρ ἡ ἀλήθεια. Diog. Laert. 9.72. Whence our "Truth lies at the bot-tom of a Well."

74. Cf. *Nov. Org.* 2.9 and 51, and *De Augm.* 3.5, where he asserts for the term Magic its proper honours.

75. Cf. *Nov. Org.* 2.63, 67, where he likens him to the Turks, whose Sultans on ascend-ing the throne murder all the seed royal. Cf. Ar. *Eth. Nic.* 1.6.1, where Aristotle declares that it is sometimes needful for truth's sake καὶ τὰ οἰκεῖα ἀναιρεῖν.

76. John 5:43.

77. Illic Pellaei proles vesana Philippi
Felix praedo jacet, terrarum vindice fato
Raptus. . . .
Nam sibi libertas unquam si redderet orbem,
Ludibrio servatus erat, non utile mundo
Editus exemplum.
Lucan. *Phars.* 10.20.

78. Tac. *Ann.* 1.3.

79. For these "four causes" see Arist. *Post. Anal.* 2.10.1. Cf. Mill's *Logic* bk. 3 ch. 5.

80. Virg. *Ecl.* 8.80.

81. See *Nov. Org.* 2.1. Datae naturae formam . . . invenire, opus et intentio est humanae scientiae. The first twenty chapters of bk. 2. of the *Nov. Org.* are an attempt at expansion of this saying.

82. Plato *Rep.* 10, init.

83. *Nov. Org.* 1.96.

84. Gen. 2:7; 1:20, 24.

85. *Nov. Org.* 2.3, efficiens et materialis causa (quae causae fluxae sunt, et nihil aliud quam *vehicula* et causae formam deferentes in aliquibus).

86. Hippoc. *Aph.* 1.

87. Eccles. 3:11.

88. *Georg.* 1.281, 282.

89. Plato *Parm.* 165, 166.

90. Perhaps Prov. 15:19, via justorum absque offendiculo.

91. Cic. *de Off.* 1.43 (154).

92. Prov. 4:12.

93. Aristot. *Phys.* 2.8.2, where he illustrates by the teeth. Also Plat. *Tim.* 3.70 and Galen *De Usu Partium.*

94. Virg. *Ecl.* 7.45.

95. For these opinions of Democritus and the Pythagoreans, see Aristot. *De Anima,* 1.2; *Met.* 1.4, 5.

96. Hor. *Odes* 2.10.3.

97. Pind. *Pyth.* 2.21.

98. Virg. *Ecl.* 10.8.

99. *Nov. Org.* 1.35. This saying of Alexander VI was called forth by the expedition of Charles VIII, which overran Italy in about five months, A.D. 1494.

100. See Ellis' note on *De Augm.* 3.4, where he suggests, most probably, that Bacon is alluding to the acts of Mahomet III, who, on becoming Sultan in A.D. 1595, put to death nineteen brothers, and ten or twelve women, supposed to be with child by his father. He adds that the practice was established as a fundamental State Law by Mahomet II.

101. *Nov. Org.* 1.45, where he calls these "eccentrics and epicycles," *lineae spirales et dracones.* Bacon was ignorant of, and incurious about, Mathematics and Astronomy at this time; and shows no good will towards Galileo and the "Copernican theory."

102. Aristot. *Phys.* 1.1.

103. Paracelsus (von Hohenheim), enthusiast and alchemist, born A.D. 1493, died A.D. 1541. He, though in a purposely obscure way, did much service to experimental philosophy.

104. Severinus, a Danish physician, died in 1602.

105. Telesius, born in 1509 at Cosenza; who, as Bacon adds in the Latin, revived the philosophy of Parmenides.

106. Fracastorius, born in 1483 at Verona; a man of greatest worth, disinterestedness, and capacity; whether as Poet, Philosopher, Physician, Astronomer, or Mathematician. But of course Bacon has no good word for him.

107. Gilbertus, Court Physician to Elizabeth and James I, a great experimentalist and discoverer in Magnetism. Bacon seems to have regarded him with especial ill-will.

108. Plat. *Alcib. Pr.* 2.124.

109. Cic. *de Orat.* 3.16, 17.

110. In his *Praenotiones.*

111. In the treatises on the History and Parts of Animals. The subject of Gesture may be said to come under the short treatises on the External Phenomena of the Animal Kingdom, and in that on the Motion of Animals.

112. Spedding gives *Basilikon Doron* bk. 3 as the place whence this quotation comes. Cf. Horace *Ars Poetica* 180, 181.

113. Qui simul cum matris affectibus compatitur, et tamen e corpore matris suo tempore excluditur. *De Augm.*

114. Diog. Laert. 2.22. Socrates speaks of a work of Heraclitus which Euripides had lent him: "Delio quopiam natatore indiget."

115. Plat. *Tim.* 69, 70 (Steph.). In the *head*, τὸ θεῖον: then below the isthmus of the neck, the mortal part of man; first τὸ μέτεχον τῆς ψυχῆς ἀνδρείας καὶ θυμοῦ; (so that Bacon is scarcely right in his censure; for neither ἀνδρεία nor θυμὸς is *anger*); then the *diaphragm* to divide the parts; then in the *heart* he placed θάρρος καὶ φόβος; and below it τὸ ἐπιθυμητικόν, ὥσπερ ἐν φάτνῃ . . . καταδεδεμένον —in the *liver*.

116. Tac. *Ann.* 16.18.

117. See Ellis and Spedding's note to *Nov. Org.* 2.48 (p. 339).

118. Virg. *Aen.* 6.747.

119. Ovid *Metam.* 1.521.

120. *Montabank*—in the old editions—from *montambanco,* a quack-doctor. Holland, in his *Plutarch,* renders the word *mount-bank.* The word was confined in meaning to a quack in Bacon's day.

121. Virg. *Aen.* 7.772.

122. Ibid. 7.11.

123. Eccles. 2:15.

124. *Buffon,* or *Pantomimus,* in the original; showing that the words were newly imported into the English tongue. The *pantomime* was then a person, not a play.

125. Ovid *Rem. Am.* 525.

126. Matt. 17:27.

127. Hippocr. *De Epidemiis.*

128. *De Medicina* 1.1.

129. Suet. *Vit. Aug.* c. 99.

130. ――― τὸν ἄκρατον
　　　　Ἔυπασεν, εἶτ᾽ Ἀΐδην ψυχρὸν ἐπεσπάσατο.
　　　　　　　　Diog. Laert. 10.15 *(Vit. Epic.).*
No *ebrius* here; *protenus* and *laetius* are suggested; but either emendation would rob the story of its point.

131. *Receipts of propriety,* i.e., proper or fit for each particular disease.

132. *Treacle and mithridatum.* In the frontispiece to the edition of Hippocrates, which I consulted, Θηριακὸν and Μιθριδατικὸν were placed side by side as the chief remedies. By treacle *(therias)* is meant, not the syrup of sugar, etc., but a composition of the parts of vipers; good for the cure of serpents' bites, and for other medicinal purposes. *Mithridate* (from king Mithridates' antidote) was a medicine of general use. "Was it not strange, a physician should decline exhibiting of Mithridate, because it was a known medicine, and famous for its cures many ages since?" Boyle's Works 2, p. 218. *Diascordium* is said to have been invented by Fracastorius.

133. By Paracelsus and his school, who were chiefly distinguished by their use of mineral medicines.

134. The passage in the Latin on the prolongation of Life, which is inserted at this point, is most curious. It was a subject to which Bacon had evidently turned his attention; for he often refers to it, and had great hopes respecting it.

135. This subject is very differently treated in the Latin. He there introduces music and painting, not as things to be repressed, but honoured.

136. Sall. *Bell. Jug.* c. 35.

137. Plat. *Tim.* 71, οἷον ἐν κατόπτρῳ δεκομένῳ τύπους, and note the observation on μαντικὴ, at the same place.

138. *Ceremonies.* The word does not now convey quite the same sense; for in these passages Bacon refers to invocation of spirits: saying (as we gather also from the Latin) that they are illicit, though used only as physical remedies without any incantation.

139. Gen. 3:19.

140. In the Latin, two desiderata are noticed; Voluntary Motion, and Sense and the Sensible: together with a curious discourse on the Form of Light.

141. Ovid *Metam.* 2.14.

142. Aristot. *Polit.* 1.5, 6; where ὄρεξις, *appetite,* is the term here rendered by *imagination.*

143. That is, Rhetoric aims at the feelings rather than at the cool judgment, and inflames Imagination till she overpowers Reason.

144. Cic. *Acad.* 4. *ad Lucullum,* 32 a. (Steph. 225). Est enim animorum ingeniorumque naturale quoddam quasi pabulum consideratio contemplatioque naturae. Or perhaps, *De Senect.* 14. Si habet aliquid tanquam *pabulum studii atque doctrinae,* nihil est otiosa senectute jucundius.

145. Numb. 11:4–6.

146. Aristot. *De Anima* 3.8.

147. Ellis and Spedding refer to Arist. *Anal. Pr.* 1.30; Mr. Markby to *Eth. Mag.* 1.1.17. Aristotle declares (*Rhet.* 1.1.1) that neither Rhetoric nor Logic has any proper subject-matter, both being purely instrumental; accordingly neither can "invent sciences."

148. *De Medicina* 1.1.

149. Not in the *Theaetetus* certainly. As Bacon in the Latin introduces the quotation with *Plato non semel innuit,* he probably is not quoting any exact passage.

150. Virg. *Aen.* 12.412.

151. Ibid. 8.698.

152. Refers, doubtless, to the rubbing of two sticks together to produce fire. Cf. *Nov. Org.* 2.2.16.

153. Τέχνη τύχην ἔστερξε, καὶ τύχη τέχνην. Arist. *Eth. Nic.* 6.4.

154. *Georg.* 1.133.

155. Cic. *p. Corn. Balb.* 20.45.

156. Virg. *Georg.* 1.145.

157. Pers. *Prol.* 8, where it is *expedivit.*

158. Virg. *Georg.* 4.1.

159. All the old editions spell the word *Issay,* and the *De Augm.* (as a genitive) *Isaï.*

160. 1 Sam. 16.

161. In the Latin, *in rebus naturalibus.*

162. This quotation is omitted in the Latin, nor can I find whence it comes; could it be a saying of Bacon's own?

163. Tesserae. Arist. *Interp.* 1.1.2—τὰ τῶν ἐν τῇ ψυχῇ παθημάτων σύμβολα.

164. Cic. *Acad.* 2.5.15. Cf. Cic. *ad Att.* 13.19.3. These very words do not occur.

165. Tac. *Ann.* 1.7.11.

166. Cic. *Acad.* 2.6.18, where κατάληψις only is mentioned. Cf. *Nov. Org.* 1.37.

167. In the Latin, Bacon explains his *experientia literata,* which treats of methods of experiment; *Venatio Panis* he also styles it. Cf. *Nov. Org.* 1.101. The *Interpretatio Naturae* is the subject matter of the *Nov. Org.*

168. Aristot. *Soph. El.* 34.

169. Matt. 13:52.

170. Cic. *Orat.* 14 (46).

171. Plato *Menon.* 80.

172. This passage is better arranged in the Latin. The paragraphs on Topics look as if they had been inserted as an afterthought.

173. Cf. Aristot. *Rhet.* 2.22.16, 17.

174. Cf. *Nov. Org.* 1.130.

175. In the Latin an inquiry *de gravi et levi* is here added as a Topic.

176. In the Latin, legitimam (Inductionis formam) ad Novum Organum remittimus.

177. Aristot. *De Motu Anim.* 3.

178. Cf. Sanderson *Logic* 3.5.

179. Sen. *Epist. Mor.* 45. Sine noxa decipiunt, quomodo praestigiatorum acetabula et calculi, in quibus fallacia ipsa delectat.

180. Cf. Plato's account of Socrates in the opening of the *Theaetetus*.

181. Arist. *Categ.*

182. This is the doctrine of "Idols," expanded in the Latin, and still more in the *Nov. Org.* 1.39–68.

183. "Idols" of the Tribe, *Nov. Org.* 1.24–31.

184. Cic. *De Nat. Deor.* 3.37.

185. Bacon's warning here is good, though his illustration was soon signally confuted by the promulgation of Kepler's laws. See *Nov. Org.* 1.45.

186. He seems to think the derivation of this term is $\mu \acute{o} \nu o \varsigma$ and $\delta \acute{\iota} \kappa \eta$.

187. Anthropomorphites, a sect which flourished in the fourth and tenth centuries; their distinctive doctrine was that as God is said to have made man in his own Image, therefore the Deity is clothed in human shape. See Mosheim *Eccles. Hist.* Cent. 10 pt. 2 ch. 5.

188. Cic. *De Nat. Deor.* 1.9.

189. "Idols" of the Cave, *Nov. Org.* 1.31–35.

190. Plato *De Rep.* lib. 7, init.

191. Aristot. *Analyt. Pr.* 2.5.1.

192. Cf. Aristot. *De Mem.* See the article in the *Encycl. Britannica*, "On Mnemonics." Cf. Cicero *De Rhet.* 3 and *De Orat.* 2.

193. *Ballerino* is Italian for a dancer.

194. Aristot. *De Interpret.* 1.2.

195. "In China et provinciis ultimi Orientis" (*De Augm.*). See a very interesting note on these paragraphs in Ellis and Spedding's edition of the *De Augm.* 6.1.

196. Aristot. *Polit.* 3.13 and Herod. 5.92. Cf. also Livy 1.54, where the story is transferred to Tarquinius Superbus. *Grandees,* in edition 1605, *grandes;* the word being not yet naturalised in the English language. According to Richardson, Burton (the Anatomy was published in 1624) spells it *grandy*. In my copy of the first edition I have not met with the word.

197. Martial. *Epig.* 9.82.

198. This quotation, which is omitted in the Latin, is only another form and application of Bacon's favourite "Antiquitas saeculi, juventus mundi."

199. In the Latin a specimen of a cipher (invented by himself when a young man at Paris) is introduced, to show how the art of writing *omnia per omnia* can be attained to. *See* also *Encycl. Brit.* verb. *Cipher.* Trithemius, Bapt. Porta, and others, wrote treatises on this art; and it is worth remembering that the Stuarts made considerable political use of it.

200. Between Ramus, whose method was one of perpetual dichotomies, and others.

201. Not so in the usual textbooks—Sanderson 3.30, 31, and Aldrich ch. 6 place it under Discourse; and it is defined as "Ratio ita disponendi partes alicujus disciplinae vel tractationis, ut facillime a nobis integra discatur."

202. I have read *use* for *ure*. For the Latin is *usus,* and the word *ure* is a rare one. Richardson's examples are all from Chaucer. The meaning of both words is the same.

203. Corresponds to the scholastic "Methodus ἀκροαματικὴ et ἐξωτερικὴ," Aldrich *Logic* 6. Bacon uses these terms in the Latin.

204. Hor. *Ars Poetica* 242.

205. This passage is equivalent to "although indeed a man would not leave some fortified place hostile to him in his rear."

206. Bacon meant here to say "diversity of Method to be used with judgment," etc.; for the Latin is "Sequitur aliud Methodi discrimen in tradendis scientiis cum judicio adhibendum."

207. *Agreeable.* "Opinionibus jampridem imbibitis et receptis affinis."

208. Arist. *Eth. Nic.* 6.3, see note in Ellis and Spedding's edition.

209. Should this not rather have been *Dichotomies*? "quam in unica sua Methodo et *Dichotomiis* obtrudendis."

210. Viz. that Propositions should be true essentially.

211. Ortelius was an Antwerper, died 1598, styled the "Ptolemaeus sui saeculi."

212. Raymundus Lully, "the Enlightened Doctor," was born in Majorca in 1225, studied Arabian philosophy, chemistry, physic, and divinity. He was stoned to death at the age of eighty in Mauretania, for preaching the gospel. For a brief account of his Method, see note to Ellis and Spedding's *De Augm.* 6.2 (p. 669).

213. Exod. 4:16.

214. Prov. 16:21.

215. Plat. *Gorg.* 462, seq.

216. Thucyd. 3.42.

217. Plat. *Phaedr.* 250.

218. Ovid *Metam.* 7.20.

219. Arist. *Rhet.* 1.1.14.

220. Arist. *Rhet.* 1.2, 7.

221. Virg. *Ecl.* 8.56.

222. *Attendances.* "Pertinent omnia ad promptuarium."

223. These were published in 1597, at the end of the volume of Essays. They are reproduced in the corresponding place of the Latin. See Arist. *Top.* 1.12.

224. Hor. *Ep.* 2.2.11.

225. Prov. 20:14.

226. Virg. *Aen.* 2.104.

227. Acts 9:25.

228. Cic. *De Or.* 1.33.

229. Tacit. *Ann.* 1.22, 23.

230. Prov. 4:23.

231. Arist. *Eth. Nic.* 2.1. *Eud. Eth.* 1.3.1.

232. Sen. *ad Lucilium* Ep. 52.

233. Demosth. *Olynth.* B, ad fin.

234. *Georg.* 3.289.

235. That is, stood to the Heathen in the place of Divinity. "Quae ethnicis instar Theologiae erant."

236. *Rhet.* 2.12.8.

237. Sen. *ad Lucilium* Ep. 53.

238. Arist. *Eth. Nic.* 10.6–8.

239. Plut. *Vit. Pomp.*

240. Rom. 9:3.

241. Cic. *Tusc. Quaest.* 5.3, of *Leo,* tyrant of Phlius, not of *Hiero.*

242. Ps. 116:15.

243. Exod. 24.

244. Jude 14.

245. Anabaptists. Bacon here refers to the doctrines held by the German Anabaptists. They believed themselves to be under special and divine influences, and therefore had no need of magistracies, of distinct ranks of men, or of restrictions in marriage.

246. Guicciardini 6.2.

247. Prov. 15:15.

248. Arist. *Rhet.* 1.5.10.

249. Diog. Laert. *Vita Diogenis,* see Ellis and Spedding's edition in loco.

250. Acts 20:35.

251. Sen. *Nat. Quaest.* 2.59.

252. Prov. 27:1.

253. Rev. 14:13.

254. Sen. *ad Lucil.* Epist. 77.

255. Sen. *ad Lucil.* Epist. 95, where the words "languida et" are wanting.

256. Virg. *Aen.* 6.730.

257. Plat. *Gorg.* 492, 494.

258. Juv. *Sat.* 10.358.

259. "Nubecula aliqua aut glaciecula," *De Augm.*

260. Cic. *de Orat.* 2.18.75.

261. Sc. the *Basilikon Doron.*

262. Sc. "The True Law of Free Monarchies."

263. Prov. 14:6.

264. Matt. 10:16.

265. Prov. 18:2. From the Vulgate.

266. Virg. *Aen.* 6.823.

267. See Plutarch *Life of Brutus.*

268. Plut. *Praec. Ger. Reip.* 24.

269. *Eth. Mag.* 1.1.3.

270. Cic. *pro Mur.* 30 (62).

271. Sen. *ad Lucil.* Epist. 71, where it is "de partibus vitae *omnes* deliberamus, de *tota* nemo."

272. Hippoc. *Aph.* 2.6.

273. Ps. 123:2.

274. Virg. *Aen.* 5.710. "Superanda omnis fortuna ferendo est."

275. Arist. *Eth. Nic.* 4.7.

276. Virg. *Aen.* 1.22.

277. *Eth. Nic.* 4.6.

278. Plaut. *Mil. Glor.* 3.1.39.

279. Tit. 1.12.

280. *Bell. Jug.* 113.

281. Tac. *Hist.* 1.50.

282. καταπέψαι μέγαν ὄλβον οὐκ ἐδυνάσθη. *Olym.* 1.55.

283. Ps. 62:10.

284. Arist. *Rhet.* 2.12–17.

285. Bacon here seems to refer to Solon's lines on Pisistratus. Ellis' edition quotes Cic. *pro Cluent.* 49.

286. Arist. *Rhet.* 2.1–11.

287. Such as Plutarch's and Seneca's.

288. *Eth. Nic.* 2.1.2.

289. Editions 1605 and 1624 have *binding*.

290. *Eth. Nic.* 2.9.5.

291. Probably St. Augustine.

292. *Eth. Nic.* 1.3.5.

293. Senec. *Herc. Fur.* 251.

294. Juv. *Sat.* 13.105.

295. Machiav. *Disc. sopra T. Livio,* 1.10.

296. Cic. *pro Mur.* 29.61.

297. Arist. *Eth. Nic.* 7.1.1.

298. Bacon seems to have thought that the Panegyric was delivered after Trajan's death. He became aware of his error before the Latin was published; for he there omits the words "in his funeral oration."

299. Plin. *Paneg.* 74.

300. Col. 3:14.

301. "Not Menander by Anaxandrides—

Ἔρως σοφιστοῦ γίνεται διδάσκαλος
Σκαιοῦ πολὺ κρείττων πρὸς τὸν ἀνθρώπου βίον."

(Spedding.)

302. Xen. *Symp.,* ad init.

303. Isa. 14:14.

304. Gen. 3:5.

305. Luke 6:27, 28.

306. Ps. 145:9.

307. Demosth. *de Fals. Leg.* p. 355.

308. Virg. *Aen.* 6.894.

309. Should this be *stupidity* or *stupor*? In the Latin it is "reddat animum—non stupidum, sed voluptatis—sensum vivide retinentem."

310. Plut. *Vit. Cat.*

311. 2 Chron. 20:33.

312. Ovid *Ars Am.* 2.312.

313. Q. Cic. *de Petit. Consul.* 11.44.

314. Cic. *ad Att.* 9.12.

315. Livy 23.12.

316. Eccles. 11:4.

317. Cic. *de Orat.* 3.133, 134 (cap. 33).

318. 1 Kings 4:29.

319. Eccles. 7:21.

320. Plut. *Vit. Pomp.* c. 20.

321. Prov. 29:9.

322. Prov. 29:21.

323. Prov. 22:29.

324. Eccles. 4:15.

325. Plut. *Vit. Pomp.* and Tacit. *Ann.* 6.46. The words *vel meridianum* are omitted in the Latin, as they should be here.

326. Eccles. 10:4.

327. Eccles. 9:14, 15.

328. Prov. 15:1.

329. Prov. 15:19.

330. Eccles. 7:8.

331. Prov. 28:21.

332. Prov. 28:3.

333. Prov. 25:26.

334. Prov. 28:24.

335. Prov. 22:24.

336. Prov. 11:29.

337. Prov. 10:1.

338. Prov. 17:9.

339. Prov. 14:23.

340. Prov. 18:17.

341. Prov. 18:8.

342. Prov. 9:7.

343. Prov. 9:9.

344. Prov. 27:19.

345. Ovid *Ars Am.* 1.760.

346. That is, soar higher than others.

347. Plaut. *Trin.* 2.2.48.

348. I have not met with this. It is attributed to Appius Claudius.

349. Liv. 39.40.

350. Plutarch *Sulla* c. 6.

351. Ezek. 29:3.

352. Hab. 1.16.

353. Virg. *Aen.* 10.773.

354. Plutarch *Caesar.*

355. Mr. Spedding states that this quotation is ascribed by Cognatus to Ptolemy.

356. Ovid *Metam.* 14.113.

357. Sueton. *Vit. Aug.* c. 99.

358. Lucian. *Hermot.* 20.

359. Virg. *Aen.* 4.423.

360. Prov. 20:5.

361. Juv. *Sat.* 2.8.

362. *De Petit. Consul.* 11.44.

363. Tacit. *Ann.* 1.12.

364. Ibid. 1.52.

365. Ibid. 4.31.

366. Liv. 28.42.

367. See Mr. Spedding's note on the *De Augm. Sc.* (p. 681), where these words are quoted with context, and traced through H. Wolf's translation of Dem. *Phil.* 1— the Greek being simply ἔστι ταῦτα τὰ τὴν ἑκάστου ῥᾳθυμίαν ἐπαυξάνοντα.

368. Tacit. *Hist.* 4.39.

369. Tacit. *Ann.* 4.52. Suet. *Vit. Tib.* c. 53.

370. Hor. *Epist.* 1.18.38.

371. Q. Cic. *De Petit. Consul.* 5.17.

372. Prov. 25:3.

373. Tacit. *Ann.* 14.57. Mr. Markby notices that Tacitus speaks "of the intrigues of Tigellinus against Plautus and Sulla, by which he induced Nero to have both of them murdered. Petronius Turpilianus was put to death by Galba, solely because he had enjoyed Nero's confidence. Vid. Tacit. *Hist.* 1.6."

374. Vid. Epictet. *Enchir.* c. 4—(λούσασθαι) θέλω, καὶ τὴν ἐμαυτοῦ προαίρεσιν κατὰ φύσιν ἔχουσαν τηρῆσαι.

375. James 1:23, 24.

376. Tac. *Ann.* 1.54.

377. Sc. Caesar Borgia, son of Alexander VI. See Guicciardini 6.3.

378. Cic. *ad Att.* 9.10.

379. Tacit. *Hist.* 2.80.

380. Mr. Spedding considers that this comes from the advice given by Medius to Alexander's sycophants—Plutarch *Quomodo quis discernere* etc., c. 24.

381. Cic. *ad Heren* 4.4.

382. Ovid *Ars Am.* 2.662.

383. That is, the last which should be made use of; "impudens certe est remedium, sed tamen, etc."

384. Cic. *Brut.* 95.

385. Livy 39.40.

386. Mach. *Discorsi sopra Livio* 3.9.

387. Demosth. *Phil.* 1.51.

388. For the same price, according to the Legend, Aul. Gell. 1.19.

389. Lucan 8.486.

390. Demosth. *Phil.* 1.51.

391. Explained by the Latin "qui occasiones quae opportune incidunt non arripiunt."

392. Prov. 30:19.

393. Plutarch *Apophthegms.*

394. Cic. *ad Att.* 10.4.2.

395. *Ad Att.* 16.15.3.

396. Tacit. *Hist.* 2.38.

397. Sueton. *De illust. gramm.* § 15.

398. Tacit. *Annal.* 5.1.

399. Caes. *de Bell. Civ.* 1.30.

400. Machiav. *Disc. sopr. Liv.* 2.10.

401. Virg. *Ecl.* 9.66.

402. *Georg.* 3.284.

403. Matt. 23:23.

404. Aristot. *Rhet.* 2.13.4.

405. *Prince* c. 17, 18.

406. *Pro Reg. Deiot.* 9.25.

407. Cic. *pro* Mur. 25 (51).

408. Plut. *Lys.*—τοὺς μὲν παῖδας ἀστραγάλοις, τοὺς δὲ ἄνδρας ὅρκοις ἐξαπατᾶν.

409. Eccles. 2:11.

410. Virg. *Aen.* 9.252.

411. Job 15:35.

412. Hor. *Sat.* 2.2.79.

413. Aurel. Victor *Epit.* 1 for Augustus; for Severus, see his life by Lampridius.

414. See Ellis and Spedding on this in the *De Augm.* 8.2.

415. Matt. 6:33.

416. ὦ τλῆμον ἀρετὴ, λόγος ἄρ' ἦσθ', ἐγὼ δὲ σε,
ὡς ἔργον ἤσκουν· δ' ἄρ' ἐδούλευνς τύχῃ.
Dio. Cass. 47.49.

417. This upon Government is very differently given in the Latin; the main subject is postponed; and two *desiderata* are discussed—the question of Enlarging an Empire, and that of Universal Justice.

418. Virg. *Aen.* 6.726.

419. Vid. Pind. *Ol.* 1.55.

420. Rev. 4:6.

421. Virg. *Aen.* 4.647.

422. Virg. *Ecl.* 2.27.

423. Themistocles to Eurybiades, Plut. *Reg. et Imper. Apop.*—πάταξον μὲν οὖν, ἄκουσον δέ.

424. Vid. Gen. 18.

425. Ps. 19:1.

426. Isa. 8:20.

427. Matt. 5:44.

428. Virg. *Aen.* 1.328.

429. Ovid *Metam.* 10.330.

430. Plut. *Alexander.* 65—εὐφυεῖς μὲν αὐτῷ γεγονέναι δοκοῦσιν οἱ ἄνδρες, λίαν δὲ τοὺς νόμους αἰσχυνόμενοι βεβιωκέναι.

431. Rom. 12:1.

432. John 3:4.

433. John 16:17.

434. 1 Cor. 7:12.40.

435. Prov. 26.2.

436. Exod. 2:11–14.

437. Matt. 12:30.

438. Luke 9:50.

439. John 19:23.

440. See Ps. 45:10, 14; or it may refer to Joseph's coat of many colours— Gen. 37:3.

441. Matt. 13:29.

442. John 4:13, 14.

443. Peter Lombard received this name after writing a work entitled "The Sentences"; a summary of Theology in four Books. The object of the work was the settlement of all disputed doctrines by a collection of sentences from the Fathers. It is perhaps superfluous to add that the work has not as yet fulfilled its object. Still he deeply affected Theology, for he laid by it the foundations of the Scholastic Philosophy. He was born at the beginning of the twelfth century; Bishop of Paris 1159; died 1164.

444. Tribonian, Quaestor, Consul and Master of the Offices to Justinian. With sixteen others he compiled the Digest—promulgated it in 533.

445. Rom. 11:33.

446. 1 Cor. 13:9.

447. Exod. 33:20.

448. Prov. 8:27.

449. John 2:25.

450. Acts 15:18.

451. 1 Cor. 13:12.

452. Matt. 24:35.

453. Luke 24:5.

454. Cabalists—expounders of the Jewish Cabala, or hidden science of divine mysteries, said by the Rabbins to have been delivered to Moses with the Law.

455. Rom. 11:20.

456. Livy 9.19.

457. John 4:24.

458. Hosea 14:2.

459. 1 Sam. 15:23.

460. Isa. 43:24.

GLOSSARY

Acception acceptation, i.e., favorable acceptance.

Accommodate an adjective in use in Bacon's day, but here equivalent to a participle and similar in meaning to *adjusted.*

Adventive adventitious, accidental—from the verb *to advene,* which is also obsolete.

Affects affections—not used here with any sense of insincerity.

Ambages circuitous paths. Bacon uses the word according to its derivation— *ambe (ἀμφί), agere (ἀγειν);* "nature worketh by *ambages* and length of time."

Antipodes the dwellers on the other side of the earth—not that other side itself.

Apprompt stir up, quicken.

Aspersion sprinkling—now used chiefly, if not entirely metaphorically, in a bad sense, then frequently in a good.

> "No sweet *aspersion* shall the heavens let fall."
>
> Shakespeare, *The Tempest,* IV.1.

Attend accompany—used actively: "I have attended them with observations."

Bird-witted incapable of continuous attention.

Blanch (also *blench*) avoid or evade.

> "And hold you ever to our special drift;
> Though sometimes you do *blench* from this to that,
> As cause doth minister."
>
> Shakespeare, *Measure for Measure,* IV.5.

Briber a receiver (not a giver) of bribes. This agrees with the origin of the word—*be-reaver,* or *be-robber,* the original usage being "thief." So, as the word changed meaning, it became "a receiver of unlawful gain." Then in modern English it is transferred to the agent who bribes.

Caption quibbling and deceit—used in fallacies. Present-day "captious" still carries this meaning.

Carnosities fleshy growths (medieval), e.g., wens.

Cautels tricks and frauds—the word having drifted away from *cautela,* while *caution* has continued to represent the Latin.

> "And now no soil, nor *cautel* doth besmirch
> The virtue of his will."
>
> Shakespeare, *Hamlet,* I.3.

Cease put a stop to—used transitively: "to cease progression."

Censured to be kept under censorship—not "blamed."

Ceremonies superstitious practices intended to invoke the aid of spirits.

> "For he is superstitious grown of late,
> Quite from the main opinion he held once
> Of fantasy, of dreams, and *ceremonies.*"
>
> Shakespeare, *Julius Caesar,* II.2.

Civil estate condition as member of a *civitas.* The highest use of the term is now almost, if not entirely, gone.

Champain open country—locus *campestris:* in Bacon's day both a substantive and (as used by Bacon) an adjective.

Circumfer (almost) transfer—a rare verb, though its sense is plain enough, and its derivative common.

Coarctation restraint.

Coevals coincident in point of time—used as a substantive.

Colliquation melting—as opposed to "coagulation."

Columbine dove-like—the innocency of the dove as opposed to "serpentine wisdom."

Compass a pair of compasses, i.e., the V-shaped instrument used for drawing circles and arcs and for taking measurements. By the change of use we distinguish between this instrument and the mariner's compass.

Compass-reaches roundabout steps taken toward the accomplishment of an object.

Complexion (probably) temperament or disposition. The word has now been degraded from the inward parts of a thing or person to the tint of the outward countenance.

Confectionary the maker of confections, not the confections made.

> "He will take your daughters to be *confectionaries.*"
>
> 1 Sam. 8:13

The word "confection" is not rightly limited to sweet stuff. Bacon uses it as equal to "apothecary," a word formed in the same manner; in Medieval Latin the apothecary was *confectionarius. Comfit* is derived from the same source.

Conscient conscious.

Consist in depend upon.

Contentation contentment.

Contestations contests, contentions.

Continent the whole extent of anything. So "the continent of Nature" is "all that comes within the limits of Nature."

Contristation trouble or distress.

Copie plenty—a French word imported into England in the sixteenth century. We still retain its adjective "copious"—and "copy" is really another form of the same word, though its usage is different. To copy is "*copiam facere exscribendi*," and perhaps carries us back to the days of the multiplication of copies of books by the hand. There is a curious use of the word in p. 169—"howsoever [a man] *change copy,* he can no ways quit himself well of it" (of contending with a fool). The Latin simply has "quocunque nos vertamus."

Corroborate strong, matured.

Decarded discarded—*de-* or *dis- carta,* to throw away one's hand at cards.

Deducements deductions.

Designments intentions.

Destituted abandoned.

Devote devoted (not *devout,* as one ed. reads it), given up to any matter—then (especially) to the worship of God.

Difficile difficult (from the French).

Digladiation hair-splitting—(properly) fencing with swords, thence with any sharp instrument (e.g., the tongue).

Dilatation power of expansion. Bacon, in saying that God is "holy in the *description* or *dilatation* of His works," seems to use these words as synonyms, whereas they are more properly used thus: "dilatation," of the expansion of the thing itself; "description," of the limitation of the thing by investigators.

Discourse illustration or investigation of a subject. So, "*discourse* of government," "discourse of business," and "discourse of reason" are all phrases used by Bacon in the original sense of the word, springing out of the Latin *discursus*—the passing from thought to thought, subject to subject; or, as in logic, from premise to conclusion; and thence the word descends to the modern usage—of discussion by talk. There is a curious usage of *discoursing* in p. 93, where Bacon uses it (unless some words have been omitted) in the sense of *final* causes.

Droumy disturbed, troubled, muddy; "to fish in *droumy* waters." The Latin has "in aquis *turbidis* piscari."

Dulceness sweetness. "Dulcet" and the verb "to dulce" are not uncommon in old writers.

Easiliest most easily.

Elench refutation of an argument or position—a technical term.

Enablement aid or means.

Estuation heat and commotion.

Exceed pass beyond the bounds of moderation—used without a case after it.

Expulsed expelled.

Exquisite carefully sought out (not "refined," as now).

Extern foreign or outward.

Extirper extirpator, i.e., "one who destroys"—the old verb being "to extirp," not "to extirpate."

Factures fashion or features of a thing, for the word "feature" is only another form of the word "facture."

Fantastical false—based upon the fancy alone, without any basis of fact or truth.

Flexuous bending and pliant.

Fripper broker. We retain the word in our "frippery"—from "frivolus," a seller of frivolous or worthless goods.

Gamester player—not with the slightest sense of *gambling*.

> "Sirrah, young *gamester,* your father was a fool."
>
> Shakespeare, *The Taming of the Shrew*, II.1.
>
> "You are a merry *gamester.*"
>
> *Henry VIII,* I.4.

Gigantine gigantic, giant-like.

Gravelled stuck or set fast in gravel; then, embarrassed.

> "Nay, you were better speak first, and when you were
> *gravelled* for lack of matter, you might take
> occasion to kiss."
>
> Shakespeare, *As You Like It*, IV.1.

The word has now passed out of the original sense. "Gravel" is derived either from *glareola* or from *gravare*—the loading of ships for ballast—or from to *grave* or dig out (to grub). The first seems to be the most probable.

Ground an instrumental accompaniment in music—the metaphor being somewhat similar to that which would connect "bass" with "base" or ground-floor of anything. The *basso* part is simply the low part—as distinct from *tenore,* midway; *alto,* high; *soprano,* above all. Thus "bass" or "ground" would be the foundation on which all rests.

Holding of pertaining to.

Humour, humorous diseased. This word (Lat. *humor,* moisture) was originally used of the four "humours" of the body: blood, phlegm, choler, melancholy; it came to a morbid state of the mind arising from excess of these; and so Bacon here uses it; "the censure of *humour,* malignity and pusillanimity" where it does not mean "ill-humour" in our sense, but rather "a diseased or jaundiced condition of mind." We apply the word in medicine to a moist diseased state of the body, in common language, to good and ill humour, or a cheerful or morose condition of temper, and to a quality of mind, difficult to define—a deep, almost solemn, sense of the incongruities which coexist in the world.

> "The *humorous* [i.e., melancholy] man shall end his part in peace."
>
> Shakespeare, *Hamlet*, II.2.

Illaqueation entanglement.

Imaginant one who imagines—a good word, though not now in actual use.

Imbar bar, hinder.

Imbase to lower, degrade—almost equal to "debase."

Impertinent out of place—according to the acceptance of the word among other writers.

Imposthumation tumour or cyst formed in any part of the body by the "humours" withdrawn from the other parts.

Improficience lack or absence of progress.

Indifferent impartial. It came naturally to mean "moderate"; thence, of course, "lukewarm" and "careless." So hard is it for one who feels strongly to help being a partisan, and so rare is a really impartial or judicious spirit.

Infirm to deprive of strength. Used by Bacon as the opposite of "affirm." The method of Socrates, he says, was to "*infirm* that which was *affirmed* by another."

Influence into used in a sense of its derivation, i.e., as of one stream flowing into another.

Ingurgitation a greedy swallowing.

Inquire inquire into.

Insatisfaction absence of satisfaction, disappointment.

Insinuation bending of oneself so as to conform to the shape of a thing—not in a bad sense.

Intend attend to.

Intrinsic internal—not, as now, referring particularly to the nature of a thing.

Invent discover (*invenire*)—the wider use of the term, now limited to the productions of man's ingenuity and skill.

Jocular pertaining to "jugglery," to which form it has been contracted in course of time. The *joculator* in low Latin was the "merry-andrew," or "juggler" (*jocus*).

Jurisconsults lawyers, jurists—being the Latin word simply transferred into the English tongue.

Leese this is the old spelling of the same verb as "to lose"; akin to it are "loss," "less," "to loose." In p. 29, Bacon uses it as equivalent to "waste," or "diminish a thing"; so, too, the termination *less* comes from this verb—blame-*less*, etc.

Levant, the the East, the Orient—not only those countries bordering the eastern Mediterranean Sea.

Lidger legate.

Liker more likely.

Limned illuminated—the derivation being the same.

Lust used by Bacon of Poesy, which is "as a plant that cometh of the *lust* of the earth"—so used as nearly equivalent to its German meaning.

Machination machine.

Magistral dogmatic.

Magistralities *magistery* was a term used by chemists for special preparations; but in p. 109, Bacon uses it as almost equivalent to "dogmatism."

Maniable manageable, tractable (through the French *manier*, from *manus*, "hand"). This French form of the word never took root.

Manured worked by hand or cultivated—first land, then intellects. This is the same word as "manoeuvre"—*oeuvre* into *ure.*

Mirabilaries works containing marvelous things.

Morigeration complaisance or compliance; "morem gerere alicui," to humour him.

Non-promovent not advancing (as are arguments *in circulo*). This is not Latin, as one edition seems to make it, but an English word, formed after the type of such compounds as "non-proficient" and "non-conforming." Bacon himself interprets it as "incurring into themselves."

Occupate occupy—also used as an adjective, "occupied."

Painful painstaking, industrious—here and elsewhere an epithet of the clergy.

Palliate palliated, mitigated.

Parcel part.

Participles partaking of more kinds than one—used generally and not solely of grammar.

Pasquil pasquinade or lampoon (from an image at Rome, to which libels and satires were affixed).

Pedantes this word was written thus by Bacon as a foreign word (Italian or Spanish, probably the latter) newly introduced into English and not acclimatized. It does not seem to carry its modern notion of "affectation" joined with learning, in the use Bacon makes of the word "pedantical" p. 141.

Percase perchance.

Ply purpose; "can bring about occasions to their ply," i.e., "can bend circumstances to their purpose."

Popularity populousness. Sir T. Browne uses "populosity," which, ugly as it is, would be the more correct form of the word.

Pragmatical officious, busy—now solely "priggish," a word that perhaps comes from it.

Prenotion a subdivision of that part of the human science which treats of the sympathy between the mind and body. Also, p. 136, the process of marking off beforehand what has no connection with the subject, used by Bacon as one of the two "intentions" or means in the received Arts of Memory.

Preposterous in the wrong order.

Presention presentiment, or previous perception inwardly of that which is about to occur.

Proficience a making of progress. "Profit" is the same word under another form.

Propriety property (in its *logical* sense).

Punctual to a point—thence, exact even to littleness and later confined to time only, in the sense of "accurate."

Punto (Spanish) ceremony, punctilio—another example of the Spanish connection with England about this period of our history.

Purgament that which purges or cleanses.

Quit acquit. So in the Bible, A.V.—"Quit you like men." 1 Cor. 16:13.

Redargution refutation.

Re-edify rebuild—the verb "edify" being used in its original signification, as "edifice" still is.

Regiment rule, government (regimen).

Reintegrate re-establish anew—not merely to renew, but to go back to the beginning, as Bacon uses it of the term *magia* which he proposes to "revive and reintegrate," i.e., "to bring back to its original sense."

Reluctation resistance. We still use "reluctant" of one struggling against what one dislikes yet is driven to.

Remora (remorare, mora) a little fish, as was thought, which, clinging to a ship's keel, stayed her course. Thence metaphor of any hindrance.

> "All sodainely there clove unto her keele
> A little fish, that men call *Remora*,
> Which stopt her course, and held her by the heele,
> That winde nor tide could move her thence away."
> Spenser, *Visions of the World's Vanitie*

Remove removal.

Respective appropriate, (almost) respectful—more exactly, having due respect or regard to the worth of the person dealt with. The honour which would be *respective* to a king would scarcely be *respective* to a squire.

Sad grave, firm, and fixed—derived from the A.S. *foet,* so that "sad" is that which is *set, fixed*; then *grave, sedate;* then *serious, mournful.*

Sapience wisdom.

Scholastical pedantic—not necessarily in a bad sense.

Seen, to be well to be esteemed.

Segregate separated part from part—as opposed to "congregate" or "aggregate."

Sever, to to be disjoined or dissevered: "seldom meet, and commonly sever."

Slug, to to render *sluggish. Slug* is from the same root as *slow.*

Solute loose and unrestrained.

Sortable agreeable to, corresponding with.

Spial spy. Shakespeare uses "espial" in *Hamlet,* III.1.

Spinosity prickliness—as of thorns.

Statua statue. The English form was in use long before Bacon's time, so that he might as well have written it instead of disfiguring his text with an unnecessary Latin word.

Stonds difficulties. It seems to be more like the joints and divisions of the stem of a plant.

Suppeditation support and supply.

Surd (almost) absurd—i.e., without proper significance; "idolatry and magic, that are full of non-significants and *surd* characters." So in mathematics, *surds* are "roots incapable of being exhibited in a finite form," and incommensurable.

Syntax arrangement in relation to one another. Bacon uses it of the "order of pursuit" in which studies may be undertaken.

Table picture (*tableau*).

Tax, to (almost) depreciated—used absolutely; "the impostor is prized, and the man of virtue *taxed*."

Terrene earthly.

Theory investigation—chiefly of things abstract, used in the original sense of θεωρία.

Thwart perverse, twisted. The verb *to thwart* is in general use, but the adjective has now disappeared. The substantive *thwart* of a boat (*cross* piece of board whereon the rowers sit) and *athwart* are also in use. A.S. *thweorian*, to wrest; *thweort*, past participle.

> "And be a *thwart* disnatured torment to her!"
>
> Shakespeare, *King Lear*, I.4.

Tractates treatises. We now have cut the word down to *tract* and its meaning down to a flimsy or short paper of a few pages. A *tractate* was a complete work on some special subject.

Trivial common and well-known—not "worthless" but (according to its derivation) "of things in the high-way," "beaten down by many feet." The sense *worthless* is later.

Typocosmy a figure or representation of the world; κόσμον τύλος.

Undervalue deficiency in worth; "what defects and *undervalues* I find in such particular acts." The verb is common enough; the substantive is not now in use.

Unperfect imperfect.

Unproper improper.

Vastness (vastitudo) a waste or desert—following the derivation of the word.

Ventosity windiness or lightness—as of air.

Vermiculate Bacon is drawing a comparison between the corruption of some solid substances into *worms*, and the tendency of sound knowledge to putrefy into idle and unwholesome "and, as I may term them, *vermiculate* questions"; where the word clearly signifies questions that are corruptions of knowledge, though some notion of entanglement and intricacy may possibly also enter in.

Vollies of wits flights (as of birds) of men of learning and wisdom. This sense is rare, if not peculiar to Bacon.

Voluble, volubility, (volubilis) apt or easy to roll—"*voluble* with the wheels of Fortune." *Volubility* is used by Bacon as an epithet of the serpent. Now used chiefly, if not entirely, of speech, and that too in rather a disparaging sense.

Whiffler piper—connected with *whiff*, a slight breath of wind; also perhaps with *waft*—such a current of air as may be made by the *waving* of a fan, though Shakespeare uses it thus:

> "The deep-mouth'd sea,
> Which, like a mighty *whiffler* 'fore the king,
> Seems to prepare his way."
>
> Shakespeare, *Henry V,* V. Chorus.

GLOSSARY OF QUOTATIONS
NOT TRANSLATED IN THE TEXT

No attempt has been made to check the accuracy of Bacon's quotations or that of the editor's footnote references. The Latin and Greek are simply translated (or, when more convenient, paraphrased) as they occur in the text, for the benefit of readers who know neither language. Numbers below refer to pages.

4   **Augusto profluens.** Proceeding from Augustus, as befitting a prince, it was "eloquence."

6   **Scientia inflat.** Knowledge puffeth up.

8   **Lumen siccum.** A dry light.
Lumen siccum optima anima (αὐγὴ ξηρὴ ψυχὴ σοφωτάτη). A dry light is the best soul.
Lumen madidum. A moist or crumpled light.

10   **Tu regere.** Do thou, Roman, remember to govern the nations—these shall be thine arts . . .

14   **Quidam tam sunt.** Some men are so wrapped up in study that they reckon anything straightforward to be complicated.

16   **Caeterum aut me amor.** Either I am deluded by love of my subject, or else there never was a State greater, holier, or richer in good examples; nor one which luxury entered so late; nor again one in which poverty and thrift were held for so long in such high honour.

17   **Verum haec.** But all these evils, together with the honour paid to wealth, will cease when public office and other things commonly desired are no longer for sale.
Qui festinat. He that hasteth after riches shall not be guiltless.
Eo ipso. They were conspicuous by the very fact of their being invisible.

18   **Quo meliores.** They are worse [men] in proportion as they are better [Jesuits].
Talis quum sis. Seeing what you are, I could wish you one of us.
Non ad vetera. Not harking back to ancient institutions which have long since been brought into contempt by debased morality.

19   **Cato optime sentit.** Cato has some excellent ideas; nevertheless he does harm to the State, because he talks as though he lived in Plato's Republic and not in the filth of Rome.
Isti ipsi praeceptores. The very teachers and professors of virtue ap-

pear to have extended its frontiers a little farther than nature intended, so that, when our minds reach the limit, we may fairly call a halt.
Monitis sum. I am unequal to my own admonitions.

20 **Satis magnum.** We are a sufficiently large theatre for one another.

23 **Execrabilis ista.** This mob is hateful that knoweth not the law.

24 **Decem annos.** I have spent ten years reading Cicero.
ὄνε [Gk.], **asine** [Lat.]. You ass!
Secundum majus. More or less [adverbially].

25 **Nil sacri es.** There is nothing sacred about you.
Devita profanas. Avoid profane and vain babblings and oppositions of science, falsely so called.

26 **Verborum minutiis.** He destroys the balance of things with the minutiae of words.
Quaestionum minutiis. They break up the bulk of knowledge with pettifogging questions.
Candida succinctam. Girt about her white loins with howling monsters.

27 **Verba ista.** These are the words of futile old men.
Percontatorem fugito. Flee the inquisitive man, for he is a prattler.
Fingunt simul. They promptly believe whatever they imagine.

30 **Oportet discentem.** The learner must believe [what he is told].
Oportet edoctum. He who has been taught should judge for himself.
State super vias. Stand ye in the ancient ways and see which is the straight and good way, and walk therein.
Antiquitas saeculi. The antiquity of the age is the youth of the world.

31 **Nil aliud.** Nothing other than having dared to despise vain things.

32 **Hic ab arte.** This man stuck fast by his own art.
Qui respiciunt. They who have few things in view express an opinion about something quite simple.

33 **Nil tam metuens.** Fearing nothing so much as that he might seem in doubt upon any subject.

34 **Declinat cursus.** She turns aside and picks up the golden ball.
Fidelia vulnera. Faithful are the wounds of a friend, but the kisses of an enemy are deceitful.

37 **Qui extendit.** Who stretcheth out the north over empty space and hangeth the earth upon nothing.
Spiritus ejus. His spirit hath adorned the heavens, and the winding serpent hath been brought forth by his guiding hand.
Nunquid conjungere. Canst thou bind together the bright stars of the Pleiades, or dissolve the circuit of the Bear?
Qui facit Arcturum. Who maketh the Bear, and Orion, and the Hyades, and the chambers of the south?

38 **Annon sicut lac.** Hast thou not stirred me like milk, and curdled me like cheese?

Habet argentum. Silver has its mine; and there is a place wherein gold is melted down, iron is taken from the earth, and brass is melted out of stone.

42 **Neque semper.** Nor doth Apollo for ever draw his bow.

Telis, Phoebe. Avenge our [the Greeks'] tears, Phoebus, with thine arrows.

45 **Quomodo Augustus.** As was Augustus, so also shall be Antoninus.

47 **Plus erat.** His unwillingness to accept was more than the other's power to give.

50 **Verba sapientium.** The words of the wise are as goads, and as nails driven in deep.

Non Rex sum. I am not King, but Caesar.

52 **Scilicet ingenuas.** Careful application to the liberal arts inevitably softens men's manners.

Nil novi. There is nothing new upon earth.

53 **Felix, qui potuit.** Happy the man who has been able to learn the true causes of events, and who has trodden underfoot all fears, inexorable fate, and the noise of greedy Acheron.

54 **Victorque volentes.** Victor, he gives laws to willing peoples and follows the road to Olympus.

57 **Occidat matrem.** Let him kill his mother, let him only reign.

Qui vetulam. Who preferred his aged wife to immortality.

Justificata est. Wisdom is justified by her offspring.

62 **Claudus in via.** A lame man on the road turns the runner aside.

Principio sedes. First you must find a suitable place for your bees, one sheltered from the wind, etc.

63 **Difficile non aliquem.** It is difficult to include everyone, ungrateful to omit anyone.

64 **Et patrum.** Weakling children will proclaim the fasts of their parents [i.e. Ill-fed parents will produce weakling children].

66 **Hoc quemadmodum.** I have a few ideas as to how this can be done, and many can be devised; please be so good as to think over the matter.

68 **Nam qui erranti.** For he who courteously directs one who has lost his way . . .

Dicit piger. The sluggard saith: "There is a lion in the way."

Possunt quia posse. They can because they think they can.

72 **Tanquam tabula.** Like a picture of a shipwreck.

73 **Caput inter nubila.** She hides her head among the clouds.

74 **Curiosus in aliena republica.** Busied with the affairs of a foreign state.

75 **Febris ephemera.** A passing indisposition.
Antiquam exquirite. Seek out your mother of old.
Animi nil magnae. Minds wanting no high praise.

76 **Non prius laudes.** No sooner have we scorned praise than we cease to do things worthy of praise.
Memoria justi. The memory of the just is blessed, but the name of the wicked shall rot.
Felicis memoriae. Of happy, pious, and good memory.
Bona fama. Good reputation is the rightful possession of the dead.
Cum ex dignitate. Since the dignity of the Roman people requires that we commit notable events to annals but such matters as these [the erection of new buildings] to day-journals.

77 **Nosque ubi primus equis.** Where the rising Sun first breathed on us with his panting steeds, there roseate Evening lights her tardy lamps.
Demens qui nimbos. [And mimic]—fool that he was—the storm-cloud, the thunder that no one may match.

78 **Plurimi pertransibunt.** Many shall run to and fro, and knowledge shall be increased.

80 **Pictoribus.** Painters and poets [have still been allowed their pencils].

82 **Illam terra parens.** Mother Earth, so they say, enraged with the gods, was delivered of her [Rumour] last, a sister to Coeus the Titan and his brother Enceladus.

84 **Si inaequalibus.** If you add equals to unequals the results will be unequals.
Quae in eodem. Things equal to the same thing are equal to one another.
Omnia mutantur. All things change, nothing perishes.

85 **Didici quod omnia.** I have learned that all the works which God hath wrought will continue for ever; we cannot add anything to them or take anything away therefrom.
Splendet tremulo. Under her [the moon's] quivering light the sea glitters.
Omnes Caelicolas. All the denizens of heaven, all those who dwell on high.

86 **Da fidei.** Leave to faith the things that are of faith.

88 **Veni in nomine Patris.** I am come in my Father's name and ye receive me not; if another shall come in his own name, him ye will receive.

89 **Felix terrarum [doctrinae] praedo.** Glorious plunderer of the earth [learning].
Eadem Magistratuum vocabula. The same official terminology.

90 **Limus ut hic.** One hardens like clay, and the other melts like wax in one and the same fire.
De mundo. Concerning the world, concerning the universe.

91 **Formavit hominem.** [The Lord God] formed man of the slime of the earth and breathed into his face the breath of life.

Producant aquae. Let the waters bring forth, let the earth bring forth.

92 **Vita brevis, ars longa.** Life is short, the art [of medicine] is a long business.

Opus quod operatur. The work that God hath done from the beginning even to the end.

Ter sunt conati. Thrice they attempted to heap Ossa on Pelion, and then to pile leafy Olympus upon Ossa.

Sancte! Holy! Holy! Holy!

93 **Latae undique.** Broad everywhere are the ways for the wise.

Rerum divinarum. The knowledge of things divine and human.

Non arctabuntur. Thy steps shall not be straitened, and if thou runnest thou shalt not stumble.

96 **Premendo littus.** Hugging the rugged coastline.

98 **Non canimus.** We sing not to deaf ears: all woodland creatures answer.

104 **Eruditus luxus.** Refined excess.

105 **Purumque reliquit.** Left undefiled the ethereal sense and the fire of essential breath.

Ipse repertorum. [The father almighty] himself with his thunder thrust down to the waters of Styx the finder of healing so potent, this son of Apollo.

Dives inaccessos. [Circe] the Sun's rich daughter [awoke] the woods that none may enter.

106 **Et quoniam variant.** And because sicknesses vary we shall vary the art [of healing]; a thousand kinds of ill, a thousand remedies.

109 **Hinc Stygias.** Drunk, he lapped thence the Stygian waters.

112 **O urbem venalem.** Oh, if this venal city, soon to perish, could find a buyer!

114 **In sudore.** In the sweat of thy brow shalt thou eat thy bread.

115 **Ad ollas carnium.** To the flesh-pots.

116 **Cuique in sua arte.** One should trust every artist within the limits of his own particular art.

Dictamnum genitrix. His mother plucked upon Ida in Crete a dittany stalk, with down clothing its leaves and a crimson flower: a plant well known to wild goats, which eat it when fluttering arrows have lodged in their flesh.

117 **Omnigenumque Deum.** Monstrous and motley gods, such as barking Anubis, against Neptune, Venus, Minerva, etc.

Ut varias. That repeated practice might gradually evolve the various arts.

Usus uni rei. The continual practice of a single thing often surpasses both nature and art.

Labor omnia vincit. Hard work overcomes all obstacles, so does stern want in difficult circumstances.

Quis psittaco. Who taught the parrot to say "Hullo!"?

118 **Aerei mellis.** Heavenly gifts of honey from on high.

119 **Quae assensum.** That which produces assent need labour no more.

Scientiam dissimulando. He feigned knowledge by dissimulation.

Acatalepsia. Incomprehensibility.

122 **Ars inveniendi.** The art of discovery advances with invention.

126 **Loquendum ut vulgus.** One should speak like ordinary folk, but think like wise men.

130 **Coenae fercula.** I would prefer that the dishes at my dinner party pleased the guests rather than the cooks.

Quod tempore. Curiously enough, that which seems ancient in time is the last word in novelty.

132 **Via deserta.** An abandoned and impassable road.

Secundum majus. More or less [adverbially].

133 **Tantum series.** So much good method and connection may improve the common and the plainest things.

135 **Καθόλου πρῶτον.** First of the whole in general.

136 **Sapiens corde.** The wise in heart shall be called prudent; and the sweetness of the lips increaseth learning.

138 **Video meliora.** I see and approve better things; I follow a meaner way.

Orpheus in sylvis. Orpheus in the woods, Arion among the dolphins.

139 **Quod laudator.** *Sophism.* What is praised is good, what is censured is evil. *Refutation.* He praises venal men who wishes to avoid having to pay them.

Malum est. "It is bad, it is bad," saith the buyer; but when he hath gone his way, then he boasteth.

Hoc Ithacus velit. If Odysseus will this, the Atridae will be well repaid.

140 **Non est interpretatio.** *For the words of the law.* Depart from the letter of the law and you have not interpretation, but divination: when the letter is abandoned the judge is transformed into a legislator. *For the meaning of the law.* The meaning of the law in any given case is to be gathered from all the words.

Demissus est per sportam. He was let down in a basket.

Demissus est per portam. He was sent out through a door.

143 **Ante omnia.** Before all things, my son, keep thy heart; for from it there proceed the actions of life.

144 **Nocet illis.** Eloquence is harmful to those in whom it begets a desire not of something else, but of itself.

Quae si feceritis. If you will do this, you will congratulate not only the speaker here and now, but also yourselves before very long because of the improved state of your affairs.

Nec sum animi. And I well know how hard it is to make mere words avail to broaden out so strait a theme in song.

145 **Vere magnum.** Truly it is a great thing to have the frailty of man and the security of God.

146 **Necesse est.** It is necessary that I go, not that I live.

147 **Pretiosa.** Precious in the sight of the Lord is the death of his saints.

148 **In praecipitio.** On a precipice.

E tela crassiore. Of tougher material.

149 **Beatius est.** It is more blessed to give than to receive.

Magni aestimamus. We set great store upon dying late.

Ne glorieris. Boast not of tomorrow, for thou knowest not what a day may bring forth.

Opera eorum. Their works follow them.

Cogita quamdiu. Consider for how long you do the same things; food, sleep, play in regular succession. Not only a brave or wretched or prudent man may be willing to die, but even a fastidious man.

Vita sine proposito. An aimless life is dull and uncertain.

150 **Igneus est ollis.** Of fire is the force these seeds have in them, divine is the source of their being.

151 **Non uti.** Not to enjoy that you may not desire, not to desire that you may not fear, are the marks of a petty and timid mind.

Qui spatium. [A stout heart] which deems length of days the least of nature's gifts.

154 **Quaerenti derisori.** Wisdom hideth herself from a scornful seeker, but comes easily to one who is zealous for her.

155 **Non recipit.** A fool receiveth not the words of prudence, unless you say what is already in his heart.

Infelix, utcunque. [Lucius Brutus] the hapless, [shall summon] his sons to their doom.

Aliqua sunt. Some things must be done unjustly in order that many things may be justly done.

Auctorem praesentis. You have the author of present justice; you have not a surety for future justice.

156 **Necesse est.** We must now discuss virtue, its nature and whence it proceeds. For it would be practically useless to know virtue but to be ignorant of the ways and means of acquiring it. We have to ask not only what sort of thing is virtue, but also how it is bestowed. For we desire both to know the thing itself and also to become possessors of it; but this desire will not be fulfilled unless we know whence and how it is derived.

Non ita disputandi. Not so much for the sake of arguing as for the sake of living.

De partibus vitae. Everyone considers the parts of life, no one considers it as a whole.

Qui gravi morbo. Those who have a serious illness and feel no pain are mentally sick.

157 **Vincenda est omnis fortuna [Natura] ferendo.** Every stroke of fate [Every nature] must be conquered by endurance.

158 **Jam tum tenditque.** Such already was her aim and purpose.

159 **Per saltum.** By leaps, step by step.

Benignitas hujus. His affability is like that of a very young man.

Increpa eos dure. Rebuke them severely.

Cretenses semper. The Cretans are always liars, evil beasts, slow bellies.

Sed plerumque. But the temper of kings, being inclined as a rule to violence, is fickle and often self contradictory.

Solus Vespasianus. Only Vespasian was changed for the better.

Qui magnam felicitatem. Who cannot digest great good fortune.

Divitiae si affluant. If riches abound, set not your heart thereon.

162 **Tanquam aliud agendo.** As if by doing something else.

Vinum daemonum. The wine of devils.

163 **Prosperum et felix.** Profitable and undetected crime is called virtue.

Ille crucem. One man is crucified, another is crowned, as the reward of crime.

In Marco Catone. The godlike and splendid qualities we see in Marcus Cato, know that they are peculiar to him; those things which we sometimes long to possess are not given by nature but are the fruits of training.

164 **Immanitati autem.** It is fitting to oppose to bestial wickedness that heroic or divine virtue which is above human nature.

Nam ut ferae. For as neither vice nor virtue are found in a wild beast, nor are they in God; the divine nature is something superior to virtue, the bestial nature something different from vice.

Amor melior (Ἔρως σοφιστοῦ). Love is better for human life than a left-handed preceptor.

165 **Ascendam, et ero.** I will ascend and be like unto God.

Eritis sicut Dii. Ye shall be like unto gods, knowing good and evil.

Diligite inimicos. Love your enemies, do good to those who hate you, and pray for those who persecute and calumniate you, that you may be children of your Father who is in heaven, who maketh His sun to rise upon the good and the bad, and raineth upon the just and the unjust.

Optimus Maximus. Best and Greatest.

Misericordia ejus. His mercy is upon all his works.

240

Sunt geminae. Two gates has slumber; the one, so they tell us, of horn, and through it true visions may easily pass from the dead; the other of gleaming ivory, cunningly wrought, wherethrough the spirits send earthwards dreams that but lie.

166   **Sed adhuc populus.** Neither as yet had the people set their hearts unto the Lord God of their fathers.

167   **Nec vultu.** Do not let thy countenance belie thy word.

Nil interest. It is no good having an open door and a forbidding look.

Ne aut arrogans. Let me not appear arrogant or subservient; arrogance is the mask of one who forgets another's freedom, subservience that of one who is unmindful of his own.

Quid deformius. What more unseemly than to make one's life a stage?

Amici fures. Friends are the thieves of time.

168   **Qui respicit.** He that observeth the winds shall not sow; and he that regardeth the clouds shall not reap.

169   **Sed et cunctis.** Do not listen unto all words that are spoken, lest perchance thou hear thy servant curse thee.

Vir sapiens. If a wise man has a controversy with a fool, whether he be angry or laugh, he will find no rest.

Qui delicate. He that delicately bringeth up his servant from a child will later on find him stubborn.

Vidisti virum. Seest thou a man diligent in his business? He shall stand before kings, and shall not be among mean men.

170   **Vidi cunctos.** I saw all the living who walk under the sun, that they were with the youth, the second, that standeth up in his stead.

Plures adorant. More people worship the rising than the setting [or the noonday] sun.

Si spiritus. If the spirit of the ruler rise up against thee, leave not thy place; for self-control allayeth the greatest offences.

Erat civitas. There was a little city, and few men within it; and there came a great king against it and attacked it, and built great bulwarks around it, and the siege was complete. Now there was found in it a poor wise man, and he by his wisdom delivered it; yet no man remembered that same poor man.

Mollis responsio. A soft answer diminisheth wrath.

Iter pigrorum. The way of sluggards is as a hedge of thorns.

Melior est finis. Better is the end of a thing than the beginning thereof.

Qui cognoscit. The judge who hath respect of persons doth not well; he will desert truth for a piece of bread.

Vir pauper. A needy man that oppresseth the poor is like a furious rainstorm, the prelude to famine.

241

Fons turbatus. As a fountain disturbed with the foot, or a foul watercourse, so is the righteous man who giveth way before the wicked.

171 **Qui subtrahit aliquid.** Whoso robbeth his father or his mother, and saith it is no transgression, is the companion of a homicide.

Noli esse amicus. Be not the friend of a man given to anger, and walk not with a wrathful man.

Qui conturbat. He who troubleth his own house shall inherit the wind.

Filius sapiens. A wise son maketh a glad father, but a foolish son is the sorrow of his mother.

Qui celat. He that covereth a transgression seeketh friendship; but he that harpeth on a matter separateth allies.

In omni opere. In every good work there shall be profit; but where there are many words, then is often penury.

Primus in sua. He that pleadeth his cause first seemeth just, but his opponent cometh and searcheth him out.

Verba bilinguis. The words of a treacherous man are as dainty morsels, and they go down to the innermost parts of the belly.

Qui erudit. He that correcteth a scorner getteth to himself shame; and he that reproveth a wicked man getteth himself a blot.

172 **Da sapienti.** Give a wise man an opportunity and he will be yet wiser.

Quomodo in aquis. As in water face answereth face, so the hearts of men are revealed to the wise.

Qui sapit. He who understands will be able to deal with innumerable characters.

173 **Nam pol sapiens.** For truly a wise man fashions his own fortune.

Faber quisque. Every man is the architect of his own fortune.

In hoc viro. In this man there was such strength of mind and character that it seemed he would have made his own fortune no matter where he had been born.

174 **Dicis, Fluvius.** Thou sayest, "The river is mine own, and I have made it for myself."

Dextra mihi Deus. Now may they help me—my right hand, which I regard as a god, and this weapon I poise before hurling.

Caesarem portas. You carry Caesar and his fortune.

Faber quisque. Every man is the architect of his own fortune.

Sapiens dominabitur. A wise man shall rule the stars.

Invia virtuti. No road is impassable to virtue.

175 **Sola viri.** Thou alone knowest how best to approach him and when.

Consilium in corde. Counsel in the heart of a man is like deep water; but a man of understanding will draw it out.

176 **Fronti nulla fides.** Men's faces are not to be trusted.

Etenim vultu. He had guessed his annoyance from his face.

Magis in speciem. Rather for show, in flowery language, than that he should be believed to feel deeply.

Paucioribus, sed intentior. In fewer words, but more intently, and in a sincere speech.

Velut eluctantium verborum. As one of slow speech.

Solutius loquebatur. But he spoke more freely whenever he meant to be gracious.

Fraus sibi in parvis. Deceit secures for itself credibility in trifles, that it may work to better effect.

Alimenta socordiae. The nourishment of stupidity.

177   **Simul amicis.** At the same time he lavished prefectures and tribunates upon his friends.

Audita haec. These words drew from him an unusually frank observation; after chiding her with a line of Greek he warned her that she was annoyed because she did not reign.

Vino tortus. Tortured by wine and wrath.

Verior fama. A man's true character may be learned from his servants.

178   **Et hoc volo (θέλω, καὶ τὴν ἐμαυτοῦ).** I want this, and also to keep my own chosen way of life.

Et hoc volo. I want this, and also to learn something more.

179   **Alia Tiberio.** Tiberius lived a different kind of life.

180   **Sulla potuit.** Sulla could—I shall not be able to.

Omnium quae dixerat. A man who took care to draw attention to everything he had said and done.

Audacter calumniare. Throw mud boldly, some of it always sticks.

Audacter te vendita. Be loud in recommending yourself, some of what you say always sticks.

181   **Cave ne insuetus.** Be careful not to appear unfamiliar with important matters, even though this trifle is big enough to amuse you.

Saepe latet. Vice often lurks in close proximity to virtue.

182   **Idem manebat.** He remained the same, but it was not right that he should.

Versatile ingenium. An adaptable genius.

183   **Fatis accede.** Bow to the fates and to the gods.

Et quemadmodum. And just as it is accepted that the general commands the army, so also is it agreed that the state itself should be managed by prudent men; that what they think right should be carried out, and that they should not be obliged merely to follow the course of events.

184   **Qualis est via.** As is the way of a ship in the sea.

Dissimulatio errores. Dissimulation produces errors which ensnare the dissimulator himself.

Alter non recusat. The other does not refuse, but even demands that he be called what he is—a tyrant.

Ita parentis. It was right that he should thus succeed to his father's honours.

Occultior, non melior. More dissembling, not better.

Ore probo. Honest in speech, shameless in thought.

185 **Et cum artibus.** A combination of her husband's wiles and the hypocrisy of her son.

Haec omnia. He made a great business of all this.

186 **Quod nunc.** Let us attend to present business.

Sed fugit. But meanwhile irrecoverable time flies by.

187 **Haec oportet.** These ought ye to have done and not to leave the other undone.

Et ama. Love as a future enemy, and hate like a future lover.

188 **Bonae Artes.** Good arts.

Cadant amici. Let our friends fall, provided our enemies perish.

Ego si quid. If fire breaks out among my goods I will extinguish it not with water but with ruin.

Quae vobis. What kind of reward matching your merits can I think of to pay you? The first and fairest heaven itself will award you, and your own hearts.

189 **Atque affigit.** And he implanted in the soil a particle of the divine breath.

Primum quaerite regnum. Seek ye first the goods of the mind; the rest will either be present or will not perish.

Te colui. I have worshipped thee, Virtue, as a *being;* but thou art an *empty name.*

190 **Totamque infusa.** And Mind, which flows through the whole mass [of the universe], mingles with that vast body.

Et in conspectu. And before the throne there was as it were a sea of glass like unto crystal.

192 **Si nunquam fallit imago.** If the image never fails.

Verbera, sed audi. Chastise me, but listen.

193 **Caeli enarrant gloriam Dei.** The heavens show forth the glory of God.

Caeli enarrant voluntatem Dei. The heavens show forth the will of God.

Ad legem. To the law and to the testimony! If they do not according to this word, etc.

Nec vox. It is not the voice of a man.

194 **Et quod natura.** What nature provides, grudging laws reject.

196 **Quomodo possit.** How can a man be born when he is old?

244

Quid est hoc. What is this he saith to us, "A little while and you shall not see Me; and again a little while and you shall see Me"?

Ego, non dominus. I, not the Lord.

Secundum consilium. According to my counsel.

Non ego. Not I, but the Lord.

198 **O altitudo.** Oh the depth of the wisdom and knowledge of God! How incomprehensible are His judgments, and how unsearchable His ways!

Ex parte scimus. We know in part.

199 **Videmus nunc.** We see now in a glass darkly, but then face to face.

200 **Auctoris aliud.** An author must not be understood literally when referring to matters not immediately relevant to his subject.

Noli altum. Be not high minded, but fear.

203 **In spiritu.** In spirit and in truth.

Vituli labiorum. Bullocks [the offering] of [our lips].

Quasi peccatum. Rebellion is as the sin of witchcraft, and stubbornness is as idolatry.

204 **Tanquam adeps.** As the fat of sacrifice.

205 **Μὴ ταχὺς.** Do not be in a hurry to open Heraclitus' book. The Ephesian's work is very hard going. It is unrelieved darkness unless an initiate guide your steps, and then it is brighter than the sun.

αὔη ψυχὴ σοφωτάτη. A radiant soul is the wisest.

207 **Rerum pondera.** He upset the balance of fact with trifling opinions.

208 **Ἰχώρ, οἷός πέρ τε ῥέει.** Ichor, such as runs in the veins of the blessed gods.

209 **Ἔοικεν, ὦ ἄνδρες.** It appears, gentlemen, that while we were conquering Darius here [in Asia], back there in Arcadia there was a battle of mice and frogs.

Transeundum plane. One must clearly leave Vulcan for Minerva if one intends to discover the true fashion and structure of bodies.

210 **Cremutius Cordus.** Cremutius Cordus is cited as having called C. Cassius "the last of the Romans."

Postquam desiimus. After we have ceased to do things worthy of praise we think it absurd to be praised.

Vita enim mortuorum. The dead continue to live in the memory of the living.

ἵν' ἦν ζῶντες. That the glory which they acquired in life may be preserved to them in death.

211 **Illic Pellaei proles.** There lieth the mad son of Philip the Macedonian, *the glorious plunderer of the earth* (line 4 of text), swept away by avenging fate. . . . For he had been kept as a laughing-stock in case freedom

should ever restore the world to herself, he *who had set the world so useless an example* (lines 4–5 of text).

Datae naturae. The aim and purpose of human science is to discover the form of a given nature.

212 **Efficiens et materialis.** The efficient and the material cause (which are transient causes, mere *vehicles,* i.e. causes giving rise to a form in individuals).

Via justorum. The way of the just is without obstacle.

213 **τὸν ἄκρατον.** He seized the cup, then swallowed down cold Hades.

οἷον ἐν κατόπτρῳ. Like images reflected in a mirror.

214 **Est enim animorum.** For the study and contemplation of nature is, as it were, the natural nourishment of men's minds.

Τέχνη τύχην. Skill gave rise to chance, and chance to skill.

215 **Sinc noxa.** Their deceit is harmless, like the apparatus of a juggler, whose very trickery delights.

SUGGESTED READINGS

Anderson, Fulton H. *The Philosophy of Francis Bacon.* Chicago: University of Chicago Press, 1948.

Aubrey, John. *Brief Lives.* Edited by Oliver Lawson Dick. London: Secker and Warburg, 1949.

Bacon, Francis. *The History of the Reign of King Henry the Seventh. A New Edition, with Introduction, Annotation, and Interpretive Essay by Jerry Weinberger.* Ithaca: Cornell University Press, 1996.

Bowen, Catherine Drinker. *Francis Bacon: The Temper of a Man.* Boston: Little, Brown and Company, 1963.

Briggs, John C. *Francis Bacon and the Rhetoric of Nature.* Cambridge: Harvard University Press, 1989.

Church, R.W. *Bacon.* New York: Harper & Brothers, 1884.

Cohen, M. R. *Studies in Philosophy and Science.* New York: Henry Holt and Company, 1949.

Eisley, Loren. *Francis Bacon and the Modern Dilemma.* Freeport, N.Y.: Books For Libraries Press, 1970.

Farrington, Benjamin. *The Philosophy of Francis Bacon.* Liverpool: Liverpool University Press, 1964.

———. *Francis Bacon: Philosopher of Industrial Science.* New York: H. Schuman, 1949.

———. *Francis Bacon, Pioneer of Planned Science.* New York: Praeger, 1969.

Faulkner, Robert K. *Francis Bacon and the Project of Progress.* Lanham, Md.: Rowman & Littlefield Publishers, 1993.

Hesse, Mary. "Francis Bacon." In *A Critical Review of Western Philosophy.* Edited by J. O'Conner. New York: The Free Press of Glencoe, 1964.

Horton, Mary. "In Defence of Francis Bacon: A Criticism of the Critics of the Inductive Method." In *Studies in History and Philosophy of Science* 4 (August 1973): 241–278.

Jardine, Lisa. *Francis Bacon: Discovery and the Art of Discourse.* London: Cambridge University Press, 1974.

Kennington, Richard. "Bacon's Critique of Ancient Philosophy in *New Organon I.*" In *Nature and Scientific Method.* Edited by Daniel O. Dahlstrom. Washington, D.C.: Catholic University of America Press, 1991.

Lampert, Laurence. *Nietzsche and Modern Times: A Study of Bacon, Descartes, and Nietzsche.* New Haven: Yale University Press, 1993.

Macaulay, T. B. *Critical and Historical Essays.* New York: E. P. Dutton, 1961.

Perez-Ramos, Antonio. *Francis Bacon's Idea of Science and the Maker's Knowledge Tradition.* Oxford: Clarendon Press, 1988.

Quinton, Anthony. *Francis Bacon.* Oxford: Oxford University Press, 1980.

Rossi, Paolo. *Francis Bacon: From Magic to Science.* Translated by Sacha Rabinovitch. Chicago: University of Chicago Press, 1968.

Urbach, Peter. *Francis Bacon's Philosophy of Science: An Account and a Reappraisal.* LaSalle, Ill.: Open Court Publishing Company, 1987.

Vickers, Brian, ed. *Essential Articles for the Study of Francis Bacon.* Hamden, Ct.: Shoe String Press, 1968.

Weinberger, Jerry. *Science, Faith, and Politics: Francis Bacon and the Utopian Roots of the Modern Age.* Ithaca: Cornell University Press, 1985.

———. "Metaphysics and Religion: Francis's Critique of the Ancients." In *Educating the Prince: Essays in Honor of Harvey Mansfield.* Edited by Mark Blitz and William Kristol. Lanham, Md.: Rowman & Littlefield, 2000.

White, Howard B. "Francis Bacon." In *History of Political Philosophy.* Edited by Leo Strauss and Joseph Cropsey. Chicago: University of Chicago Press, 1987.

———. *Peace Among the Willows: The Political Philosophy of Francis Bacon.* The Hague: Martinus Nijhoff, 1968.

Whitney, Charles. *Francis Bacon and Modernity.* New Haven: Yale University Press, 1968.

Yates, F. A. *The Rosicrutian Enlightenment.* London: Routledge and Kegan Paul, 1972.

Zagorin, Perez. *Francis Bacon.* Princeton: Princeton University Press, 1998.

INDEX